Together with Konrad Lorenz, Niko Tinbergen is generally acknowledged as the founder of ethology. Professor Tinbergen has devoted a lifetime to exploring the behavior of many types of animals. He has also worked tirelessly for the use of scientific methods in the study of human behavior, both normal and abnormal.

This volume includes accounts of Tinbergen's remarkable laboratory experiments as well as his significant general papers. The selections examine the animal roots of human behavior, the relation of behavior and natural selection, the character of appeasement signals, and the nature of ethology. "Early Childhood Autism," written by Professor Tinbergen and his wife, is among the most important of these papers. It is a pioneer work in applied ethology and is a product of thirty years of observing non-verbal expression in both animals and children. "Functional Ethology and the Human Sciences" is also included.

The Animal in its World

By the same author

The Study of Instinct
Oxford
Social Behaviour in Animals
Methuen
Curious Naturalists
Country Life
Animal Behaviour
Time–Life
The Animal in its World
Volume One

and others, including translations into eight languages

NIKO TINBERGEN FRS
Professor of Animal Behaviour and Fellow of Wolfson College, Oxford

The Animal in its World

Explorations of an Ethologist

1932–1972

FOREWORD BY SIR PETER MEDAWAR, FRS

Volume Two

LABORATORY EXPERIMENTS
AND GENERAL PAPERS

HARVARD UNIVERSITY PRESS
CAMBRIDGE MASSACHUSETTS

ISBN 0-674-03727-8 (cloth)

ISBN 0-674-33728-6 (paper)

Library of Congress Catalog Card Number 72-94876

Second Printing 1975

PRINTED IN THE UNITED STATES OF AMERICA

Dedicated to E.A.T.
for her interest, encouragement
and tolerance

Foreword

BY SIR PETER MEDAWAR, FRS

Niko Tinbergen is one of the grand masters of Ethology, and the papers published here are among its most important documents: they are a source-book for students of animal behaviour and will give the historian of Ideas an insight into the early days of one of the most influential movements in modern science.

Anybody who thinks that Ethology consists of a passive imbibition of the information proffered by nature still has much to learn. The first stage in a behavioural analysis is, of course, to observe and record what is actually going on. This will involve intent and prolonged observation until what an untrained observer might dismiss as a sequence of unrelated behavioural performances is seen to fall into well-defined and functionally connected sequences or behaviour structures. These behaviour structures do not declare themselves in any obvious way. Their identification depends upon an imaginative conjecture on the part of the observer which further observation may or may not uphold. As in other branches of science, this is a creative process in which the imagination must take the initiative. Another important element in the ethological approach is the comparison of behaviour structures among different but related animals, which in turn opens out the possibility of identifying homologies of behaviour. The word 'homology' is not easy to define in any context, but in an ethological context it may be exemplified by saying that the behaviour associated with mother-love and suckling is obviously homologous or genetically cognate in man and in apes. Clearly this complex behavioural repertoire did not spring into being fully formed with the inception of the species *Homo sapiens*.

In Tinbergen's early days as a research worker 'experimental' was the boss word, much as 'molecular' is today. Everything had to be experimental: Embryology, Pathology, Physiology, and, if possible, the study of behaviour. Today anybody who studies 'molecular' something instead of humdrum old everyday something feels an unaccountable increase of stature. Some of the early critics of Ethology believed that, because of its mainly observational character, it could not possibly be really in the mainstream of biological research. The idea that there is something *essentially* meritorious about experimentation has been carried over from Bacon's original usage of the term[1] and the advocacy that went with it. It just so happens, however, that some of Tinbergen's earliest work was experimental.[2] Much of the success of his work is due to his adoption of a judicious blend of observation and critical experimentation.

Ethology was soon recognised as one of the really important developments in modern biology, and became the subject of excited discussion in zoological departments. Many of the brightest zoological students and natural historians cherished ambitions to study under Tinbergen in Oxford. I say 'in Oxford', although Tinbergen was Professor of Zoology in the University of Leiden, because one of Professor A. C. Hardy's many bright ideas had been to persuade Tinbergen to come to work in Oxford and enjoy the facilities and spacious academic atmosphere he felt sure Oxford could provide. By juggling with half-vacancies, as we shall soon all be juggling again, a post was created which Tinbergen agreed to take. (Harvard wooed him too, but Tinbergen has not regretted his choice.) One thing Oxford could not supply: a seashore; but a Landrover overcame the difficulty and Oxford's rough equidistance from the sea in all directions came to be seen as an advantage by making Oxford literally the centre of Herring Gull field stations.

It is a source of great pride to me that, when funds for Ethology were running low, I was able to interest the Nuffield Foundation in Tinbergen's work which they supported by grants totalling several thousand pounds over a period of ten years, whereupon the support of his research devolved upon the Natural Environment Research Council. I was relieved that the Foundation took my advice instead of that of the eminent neurophysiologist whom they also consulted and who said of Ethology, 'Why, that's just birdwatching, isn't it?'

The paper on Autism is to be read as a study in the application of ethological methods to human infants. Notice here, for example, the identification of certain elements of autistic behaviour in the behavioural repertoire of normal children, and the implied hypothesis that Autism may be a consequence of overstimulation—of too much intrusive attention perhaps, and not, as some immature psychologists have been inclined to feel, of too little attention. The picture of Tinbergen that comes to mind through this paper,

[1] Medawar, P. B., *The Art of the Soluble*, Methuen, London, 1967; and also *Induction and Intuition in Scientific Thought*, Philadelphia and London, 1969.

[2] Tinbergen, N., 'Curious Naturalists'. *Country Life*, Feltham, Middlesex, 1966; and also *The Study of Instinct*, Oxford, 1951, p. 32.

with his enthusiasm, kindness, lack of pretension and acute observation, is entirely authentic.

Tinbergen's own childhood was spent amidst a happy and cheerful family and he showed an early predilection for a naturalist's pursuits. One of his school reports, with an almost uncanny lack of prescience, said that his powers of application were not such as to equip him for a career as a natural historian. Tinbergen spent much of the war in a hostage camp and was able to reflect, not for the first time, that wild animals are less to be feared than malevolent human beings.

Tinbergen himself has rarely attempted to establish significant homologies of behaviour between human beings and lower animals, a subject on which his pupil Desmond Morris has written so skilfully and so divertingly (*The Naked Ape*, London, 1967). 'What is there', we may well ask, 'in all this naked ape business?' The approach can be an eye-opener to those who had not realised that the human behavioural repertoire is of some evolutionary depth. Human behaviour did not spring into being fully fashioned but, like human bodily structures, must have evolved from lowlier precursors. Of human bodily structures Darwin said that, in spite of what he was kind enough to call our exalted powers, 'Man still bears in his bodily frame the indelible stamp of his lowly origin.' Without doubt the same is true of many human behavioural performances. As with many of the so-called 'behavioural' sciences, the difficulty arises at a demonstrative or evidential level. If any ill-disposed critic says of some theorem 'I simply don't believe it', its champion has no recourse except to appeal to the general plausibility and reasonableness of his argument. There is no well-understood and authenticated remedy for disbelief, as there is in the conventional or 'hard' sciences.

Tinbergen has had a number of bright pupils of whom he is very proud and with whom he remains on good terms although Ethology is a controversial and rather passionate subject. Through them and through his own work he achieved the ambition he first formulated in the wartime hostage camp: to bring Ethology to the English-speaking world (in effect to England, because America already had a long tradition of animal behavioural research, although in quite a different style—more 'scientific', in what I think of as the wrong-headed usage of that term: more measuring instruments and many more measurements but less to do with natural behaviour or indeed with nature at all. The terminology that went with it—tropisms, taxes and the like—now have to me as olde-worlde a flavour as any Tea Shoppe in a small English cathedral town.)

Tinbergen sometimes describes himself as an artist *manqué* rather than as a scientist (though the two descriptions are not incompatible), but he is in fact a first-rate scientist whose austere standards of research have steered Ethology safely between the beguilements of imaginative story-telling and the barrenness of compiling inventories of facts.

Contents

References to the Original Papers

SECTION III

11 N. TINBERGEN and D. J. KUENEN (1939). 'Über die ausloesenden und die richtunggebenden Reizsituationen der Sperrbewegung von jungen Drosseln (*Turdus m. merula* L. und *T. e. ericetorum* Turton)', *Z. Tierpsychol.*, **3,** 37–60.

12 R. HOOGLAND, D. MORRIS and N. TINBERGEN (1957). 'The spines of sticklebacks (*Gasterosteus* and *Pygosteus*) as a means of defence against predators (*Perca* and *Esox*)', *Behaviour*, **10,** 207–36.

SECTION IV

13 N. TINBERGEN (1965). 'Behaviour and natural selection', in *Ideas in Modern Biology*, ed. J. A. Moore. Nat. Hist. Press, New York, 521–42.

14 N. TINBERGEN (1959). 'Einige Gedanken über Beschwichtigungs-Gebaerden', *Z. Tierpsychol.*, **16,** 651–65.

15 N. TINBERGEN (1969). 'Ethology', in *Scientific Thought 1900–1960*, ed. R. Harré. Clarendon Press, Oxford, 238–68.

16 N. TINBERGEN (1964). 'The search for animal roots of human behaviour'. Unpublished lecture.

17 N. TINBERGEN and E. A. TINBERGEN (1971). 'Early Childhood Autism—a hypothesis'. Abridged from *Z. Tierpsychol.*, Suppl. 10, 1–53.

18 N. TINBERGEN (1972). 'Functional Ethology and the human sciences', *Proc. Roy. Soc. Lond. B*, 182, 385–410.

Section III
Laboratory Experiments

Author's Notes

These two particular papers deal with subjects suitable for repetition, and (above all) elaboration, in class practicals. However, both also contain some points of topical interest. The conceptual separation between eliciting and orienting stimuli, made initially by Lorenz, is applied, with rather remarkable results, in the first paper. However unsophisticated this paper is in many respects it does contain some interesting steps from playfully modifying the natural external situations to a more systematic pinning-down of some relevant parameters. Finally, there are some methodologically sound tests bearing on the 'nature-nurture' problem, such as the forcing, upon the growing animal, of an abnormal treatment and finding that, contrary to the expectations of dyed-in-the-wood environmentalists, the selective responsiveness found in a young animal at an early stage can be extremely environment-resistant. The fact for instance that the deliberate exposure of young Blackbirds to white objects did not result in a change of preference from black to white was not only overlooked by contemporary critics but was, by one of them, actually stated to have been omitted. I call attention to this because, as happens so often in scientific controversies, champions of extreme positions in an 'either-or' issue often state their differences in exaggerated form, and so make both themselves and each other appear to be more firmly entrenched than they really are. This still applies to the nature-nurture problem in behaviour; however much both Lorenz and Lehrman for instance have mutually adjusted their views, they have still not come together; there is still a discrepancy between the differences in their theoretical attitudes and the similarity of the actual research done; both authors and their followers *do* study the developmental *process* rather than aim at a classification of innate and acquired *behaviour*; and both schools *do* study the interaction between internal and external sources of programming of the behaviour machinery.

The second paper has as its main theme a problem of survival value. It can be said to have been the first step in a study of adaptive radiation in sticklebacks. As such it naturally concerns structure as well as movement, and it considers, in ways later elaborated in so many studies of adaptive radiation, one feature (viz. the use of spines in defence against predators) as part of an integrated adapted system. Whether or not we were right in considering protection by means of spines as somehow 'primary', there seems little doubt about the truth of our suggestions that protection by means of spines, habitat selection, overall tameness, reproductive coloration and even some details of the spawning behaviour must be seen as components of one major adapted complex.

(*From the Zoological Laboratory of Leiden University*)

11

The Releasing and Directing Stimulus Situations of the Gaping Response in Young Blackbirds and Thrushes (*Turdus m. merula* L. and *T. e. ericetorum* Turton) (1938)

1. Introduction and Theoretical Approach

In young blackbirds and thrushes, the typically passerine gaping movement is an abrupt forward movement of the head with simultaneous opening of the beak, which exposes the orange-yellow lining of the mouth. The 'gape' is then maintained for a few seconds in this position and ends when the adult bird has provided food. If no food is given, gaping can continue for up to some 15 seconds. During gaping, the neck is fully extended. For the first few days of life, the head is not kept quite still; it wobbles as if it were too heavy for the neck muscles. These wobbling movements later disappear. Whilst gaping, the birds utter a high-pitched, plaintive call which becomes louder every day.

The direction in which the neck is extended is not the same throughout the whole nestling period; the young birds initially gape vertically, while later they direct the gape towards the adult's head.

Already when the birds are only about 1 week old, gaping begins to be accompanied by other movements. The nestlings get up and make flicking movements with one or both wings, and these develop gradually into flying.

Gaping of young blackbirds, like that of other passerines, is influenced by sensory stimuli in two different ways. In the first place, it must normally be *elicited* by outside stimuli. Apart from rare 'vacuum activities' (**1**) (which occur only after prolonged food deprivation), young blackbirds gape only when the adult bird arrives at the nest. Persistent begging behaviour, as is seen, for example, in herons, does not occur in blackbirds and thrushes. Secondly, the gaping movement is *oriented.* In the species investigated this orientation, as stated, alters during development.

The task we set outselves was that of analysing both the eliciting and the orienting stimulus situations by means of experiments with dummies. As far as we are aware, no study of this kind has yet been published. Shortly after we started our study, K. Lorenz's comprehensive paper on the role of the companion in the bird's environment (**2**) appeared. In the treatment of the 'parental companion', various points relevant to our problem were made, and we shall need to come back to this paper later.

Following pilot experiments in 1935, the main experiments were carried out in the spring of 1936 and 1937. One part of the experiments on visual orientation with the blackbird was refined and extended by Tinbergen during a visit to Dr K. Lorenz in Altenberg.

At this point, we thank Dr and Mrs K. Lorenz for their kind hospitality and keen interest.

2. Material and Methods

For the experiments in 1936 and 1937, we used a total of 8 blackbird broods (*merula*, 31 individuals) and 3 song-thrush broods (*ericetorum*, 11 individuals). Most of them were taken from the nest when 5 days old. All were reared with a mixed food preparation of cooked egg, worms, mealworms, and 'Sluis' Thrushfood'. In the first year (1935), the diet was apparently not quite adequate; in the fifth week of life, symptoms of disease or possibly deficiencies appeared. After they had been fed more fresh earthworms the young birds recovered, but

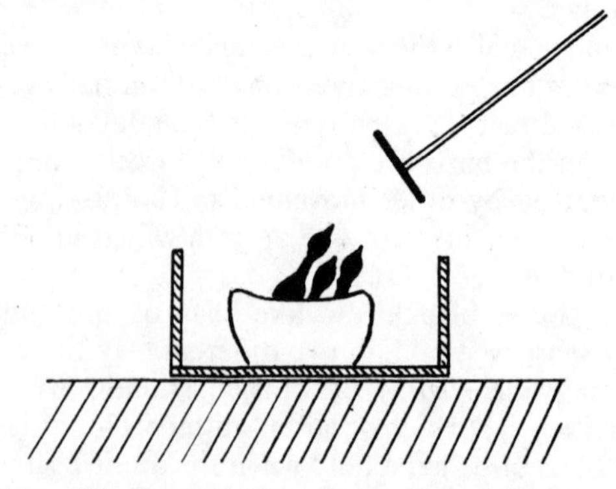

Fig. 112. Presentation of a cardboard dummy.

results obtained with these birds have not been included. We are extremely grateful to Mr J. Zwartendijk and Mr J. W. Ebbinge of Boskoop for the provision of many broods of the two species.

Each brood was kept in the original nest in a wooden box covered with glass or cardboard and fed five to ten times a day. Each feeding session was preceded by experiments as follows.

In order to identify the releasing stimuli, it was important that gaping should be a clear-cut response to the stimulus presented. When the observer approached the birds, there was usually a short bout of gaping. Because this could have been due to any aspect of the situation, this initial response was not recorded. Approximately half a minute after the gaping had ceased, the actual experiment was done, by presenting a dummy (Fig. 112) and it was the response to this dummy that was scored.

After the birds had once more settled down, the same stimulus would be presented once more, or even several times in succession. However, the stimulus was not presented too often without the birds being fed soon afterwards, and there was little or no waning in the gaping to initially effective stimuli. In addition, the possible formation of positive associations was avoided by continuous alteration of the stimuli.

When a negative result was obtained, we always presented, as controls, stimuli known to be very effective. As we were concerned only with a qualitative analysis, we did not feel it necessary to standardize the control stimulus too precisely.

With experiments on orienting stimuli, the part of the dummy to be investigated in any one case was moved after the birds had been induced to gape, so as to observe the influence of this movement on the gaping direction. Since from the tenth to the eleventh day onwards, gaping is directed at the adult's head these experiments often involved movement of an artificial 'head' with respect to a 'body'. This was best achieved by rotating a circular 'body', to which the 'head' was attached. Thus the body remained in effect stationary. Sometimes we made the birds choose between two different types of 'head'; these were presented in completely equivalent positions on the same 'body', or sometimes without the body.

In order to check the validity of observations on the captive birds, many of the experiments were repeated in the field with undisturbed broods reared by their own parents. For this, we employed a total of 5 blackbird broods (16 birds) and 3 song-thrush broods (10 birds).

Mr I. C. J. van Dusseldorp very kindly permitted us to perform the field experiments on his country estate 'Zuydwijck' near Wassenaar.

3. Elicitation

IMPACT ON THE NEST-RIM

Slight shaking movements of the nest always elicit gaping. This stimulus only fails to work immediately after feeding, when the birds are completely satiated. This was such a reliable response that we were able to use it consistently as a control for adequate response motivation in our other experiments. If the birds did not respond to a light impact on the substrate, a negative result obtained previously with another stimulus was discounted.

Only a light impact, roughly equal to that provided by an adult bird alighting on the nest, released gaping. If a stronger shaking movement was made, the nestlings frequently crouched.

SELECTED PROTOCOL RECORDS

8.5.36 *Song-thrushes, 8 days old.* Lid raised, hand moved back and forth over nest. No response. Slight impact made on nest-rim. All three immediately gape—vertically. After the birds have settled, no response is given to hand movement.

9.5.36 *The same birds.* The experiment is repeated with the same result. Then: gentle knocking or pushing against the wooden box containing the nest elicits gaping. A shaking movement of similar strength, inaudible to the human ear, similarly elicits immediate gaping. If an impact is made without previously raising the lid, so that the animals are completely in the dark, one can hear the beaks hitting the lid. Therefore light is not a necessary condition.

28.6.36 *Blackbirds, 8 days old.* Raising of the lid and removal of the faeces with forceps does not elicit gaping. A very slight impact is made on the nest; immediate vigorous response.

22.5.37 *Blackbirds, 5 days old.* Immediate vigorous response to very slight jarring of the nest. Vigorous shaking of the nest suppresses gaping.

23.5.37 *The same birds.* A heavy object is placed on the table on which the birds are situated. The impact does not elicit gaping. Immediately afterwards, light tapping on the table elicits a vigorous response. Touching of the beak margin.

Like many other passerines, blackbirds and thrushes have, in both corners of the mouth, a thick yellow swelling, which does not disappear until the birds have begun to eat independently. If this swelling is touched with an object of some kind, the gaping response frequently follows. However, this stimulus is not as effective as the shaking stimulus. Touching with wooden, glass or metal rods coated with earthworm juice or juice pressed out of the food mixture employed does not elicit any more vigorous a response than touching with uncoated glass rods. The stimulus is therefore purely tactile and not chemical.

PROTOCOL RECORDS

10.5.36 *Song-thrushes, 10 days old.* At the first feeding, when the animals—as usual—are not very responsive, a response is given only to the second impact on the nest. Touching of the gape swelling then immediately elicits gaping.

13.5.36 *The same birds.* After repeated gaping to visual stimulation, no further response is given to any visual stimulus. Touching of the corner of the mouth is immediately followed by a response.

26.6.36 *Blackbirds, 5 days old.* Touching of the beak margin is immediately followed by gaping. Similar touching of other areas of the body is not followed by a response.

9.5.36 *Song-thrushes, 9 days old.* Gaping is elicited twice by impact on the nest-rim. After the animals have settled down, touching of the beak margin produces a good response. With the smallest nestling, the response is only given after a second touch. Similar touching of the back, head or wings does not elicit a response from any of the three nestlings. Touching the corner of the mouth then produces immediate gaping. We touch the wings; no result.

AIR CURRENTS

When the adult bird settles on the nest, the nestlings are always exposed to the air turbulence caused by the alighting parent. The possible effect of this was tested with a small bellows with which air movements of many different kinds were produced. Gaping was frequently elicited, though not as often as by the stimuli already considered.

The most effective stimuli are localised and abruptly occurring puffs of wind of medium intensity. These must be directed at the nestling's head. The duration of the puff is of no importance, as is to be expected, since the birds begin to gape at once.

As long as the birds are still squatting in the nest with their eyes closed, a puff of wind can easily be presented without other effective stimuli. Later, however, when the eyes remain continuously open, we ensured that the movement of the bellows was as inconspicuous as possible; on other occasions the bellows were, as a control, moved in the same way but without being blown. In this way, we succeeded quite well in demonstrating the effectiveness of air turbulence alone.

In contrast to the stimuli already dealt with, wind is not equally effective in all phases of development. Older birds no longer responded to it. The oldest animals which still responded were 16 days old.

VARIATIONS IN LIGHT INTENSITY

According to Lorenz (**3**, p. 194) 'very many young hole-nesting birds' gape in response to darkening of their surroundings. The same might apply to the blackbirds and thrushes, which of course usually breed in concealed places where arrival of the adult bird will likewise reduce the illumination. Conversely, increase in illumination when one of the parents rises to make room for its partner could also represent an effective stimulus. However, the behaviour of adult blackbirds and thrushes on the nest (where they somehow seem to avoid each other) renders the latter *a priori* improbable.

Because darkening due to the arrival of the adult bird is under natural conditions always accompanied by movement, experimental study requires exclusion of all movement, and production of a change in overall illumination alone.

Two different procedures were followed. We either switched on and off a light-bulb fixed behind an opaque glass screen above the animals (Fig. 113) or used two light-bulbs, invisible to the birds, which illuminated a wooden board suspended above the nest (Fig. 114). The switches could be used without making a noise.

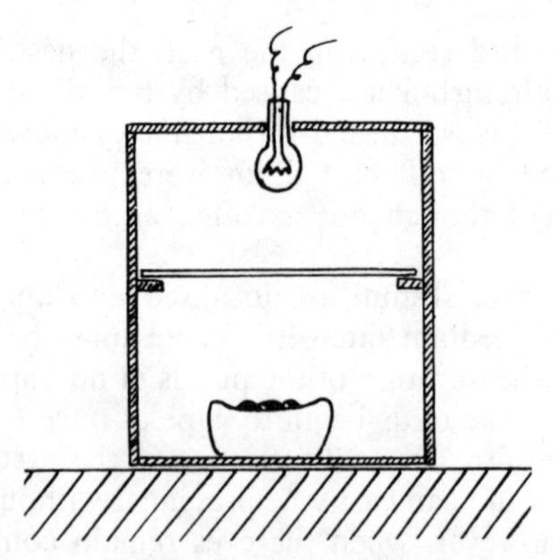

Fig. 113. Arrangement for providing abrupt variations in intensity of illumination with direct lighting.

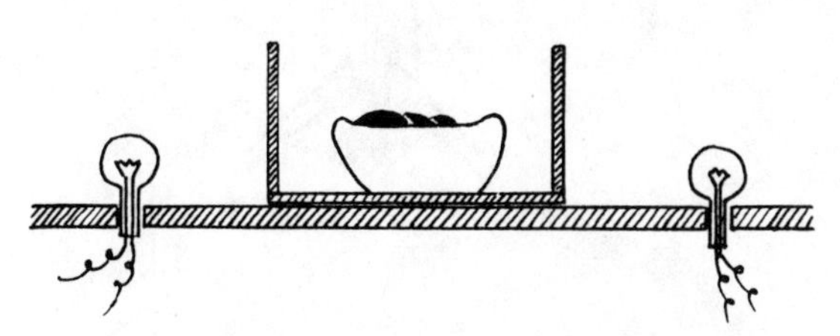

Fig. 114. Arrangement for varying reflected light.

None of the experiments produced a positive result; we never observed a gaping response to either darkening or illuminating. However, inadvertent movement often elicited gaping. In view of Lorenz's statement cited above, it would certainly be rewarding to follow up this question with a hole-nesting species, using appropriate methods.

VARIATIONS IN TEMPERATURE

Rising of the adult bird from the nest not only produces an increase in light intensity; it also causes a temperature change. Lorenz (3) postulates that an abrupt fall in temperature elicits gaping in some young passerines. In view of the behaviour of the adult birds (which do not regurgitate but leave the nest for foraging trips of some length), this stimulus also seemed unlikely to have any effect but we investigated the influence of temperature variation in a series of experiments. Heat could be radiated down onto the nest from an infrared-radiating electric heater suspended about 50 cm above the nest (Fig. 115). Abrupt insertion of a glass plate just above the nest cut off the heat radiation, whilst sudden removal of the plate produced a sudden rise in radiation. The intensity was changed by varying the distance between lamp and birds.

The result was definitely negative: the birds responded neither to a fall nor to a rise in temperature. One example from protocol records should suffice:

9.5.36 *Song-thrushes, 9 days old.* Lamp allowed to heat up, glass plate removed, no response. Impact on nest-rim; immediate gaping. After the birds have settled, we slide the glass plate over the

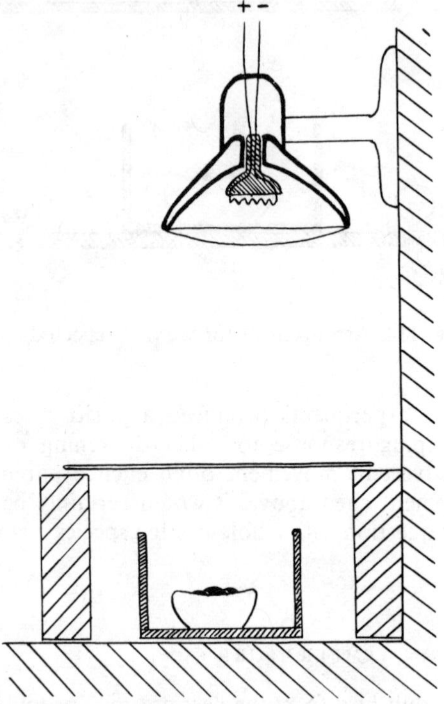

Fig. 115. Arrangement for providing abrupt temperature changes.

nest; no response. Gaping immediately given following a light impact. Glass plate removed, no response; replaced, no response. Impact, gaping once again, etc., etc.

SOUND

As is known from field observations, adult blackbirds and thrushes do not give a special feeding call. However, there are blackbird pairs in which the male goes to the nest uttering a short song strophe. With one such pair, we were able to observe that this call elicited gaping from nestlings about 10 days of age. Since the particular environment of this nest prevented the nestlings from seeing the parent when it called, the song itself must have had a releasing effect. Since none of the other young birds we investigated could be induced to gape by any sound whatsoever we assume that these particular birds had become conditioned to the sound. On innumerable occasions, we investigated the effect of a wide range of noises including the natural sounds which might function as releasers (calls of adult birds, flight sounds, etc.). Apart from one exception, these sounds had no effect. The exception was with one blackbird brood kept in the laboratory, who, surprisingly, gaped briefly and exclusively to the mild form of the blackbird's warning call ('tchuck-tchuck-tchuck-tchuck').

Otherwise, both in the blackbird and the song-thrush, any utterance of the warning call by the adults at once extinguished gaping. Even a weak imitation of the warning calls and weak noises produced by clapping hard objects together had the same effect.

This fact incidentally caused us a fair amount of difficulty with our field experiments. As soon as the adult birds gave the warning call, the nestlings crouched and it was difficult to elicit gaping even with optimal stimuli.

The blackbird brood exhibiting the abnormal gaping response to the warning call was brought to the Institute when 5 days old on 22.5.1937. After a number of non-experimental feeding sessions, some experiments with tactile stimuli were performed at 6.00 p.m. Then: they gaped vigorously in response to 'tchuck-tchuck' and to various of our vocal sounds. The same occurred later on in the evening.

24.5.37 *The same animals, 7 days old.* Gaping follows our imitation of the warning call, but not shouting or other sounds.

29.5.37 *The same animals, 12 days old.* The birds no longer gape in response to our imitation of the alarm call, but crouch. Later during the same day, we sometimes obtain gaping responses, sometimes crouching responses to the same stimulus.

From this day on these birds always responded with crouching. Thus, remarkably, they possessed selective sensitivity for the alarm call, but it was originally followed by the 'wrong' response. The later 'correction' had, we believe, nothing to do with experience, since our procedure did not provide any opportunity for conditioning in this direction. Our interpretation of this observation is that the behaviour resulted from abnormally late maturation of the normal nervous association between the alarm call as the effective sensory stimulus and crouching as the response. The peculiar fact that the alarm call initially released a different response raises a new question which cannot be answered on the basis of this single observation.

VISUAL STIMULI

As has already been said, the young first open their eyes on the 9th or 10th day of life. Up to that time, it is impossible to evoke gaping with visual stimuli of any kind. But even after visually elicited responses have appeared, the effect of visual stimuli is slight, since when resting the animals usually have their eyes closed; they open them only after tactile stimulation and then respond to visual stimuli.

SELECTED PROTOCOL RECORDS

26.5.36 *Blackbirds, 9–10 days old.* Cautious lifting and subsequent moving of the wooden lid does not evoke a response. As soon as we tap the nest-rim, the birds respond. After gaping has declined somewhat, but before the birds have closed their eyes, a hand is moved to-and-fro above the nest; this elicits immediate, vigorous gaping.
Later the same day. Tapping the closed box elicits gaping, but subsequent lifting of the lid does not. The eyes are closed. An impact produces gaping. After the response has ceased, the animals settle but keep their eyes open. A circular white cardboard disc (20 cm diam.), a similar black disc, a white wooden rod (25 × 0·75 cm), and a similar black rod are presented successively over the nest. All of these objects elicit vigorous gaping.
10.5.36 *Song-thrushes, 10 days old.* Careful removal of the lid produces no response, but a light impact does. After the nestlings have settled, a finger presented some 15 cm above the nest immediately releases gaping. The same evening, a young bird responds to the sight of the black rod without a previous tactile 'alert'. When the same rod is subsequently presented again, none of the nestlings responds. However, a control with impact produces a positive response.

After the 10th (blackbird) or 11th (song-thrush) day, the eyes of satiated, resting birds remain open for periods of increasing length. Correspondingly, the effect of visual stimulation increases. As the above protocols show, the visual stimulus situation is extremely unspecific. However, further analysis demonstrates that certain conditions must be fulfilled. The following factors were investigated: height, size, colour, shape, movement.

RELATIVE HEIGHT

To elicit a gaping response with visual stimuli, an object must always be presented above the horizontal plane passing through the two eyes. If objects which normally evoke a response are not high enough, the birds do follow the object with their eyes but never gape in response.
19.6.36 *Blackbirds, 10 days old.* A black disc is moved to-and-fro

below the nest. The nestlings, which extend their heads far outside the nest, look at the object but do not gape.

A black cardboard disc 8 cm in diameter is moved towards the nest at about eye-level. The upper edge of the disc is approximately 4 cm higher than the birds' eyes. One nestling opens its beak once, but does not gape. At the next presentation, one young bird gapes.

27.5.37 *Song-thrushes, 12 days old.* Gaping is not evoked by a rod presented below eye-level, though the rod is continuously fixated by the nestlings. An impact on the nest-rim evokes immediate gaping.

8.6.37 *Song-thrushes, 8 days old.* When a black rod is moved towards the nest at eye-level, it is fixated but there is no gaping. As soon as the rod is raised a few centimetres, immediate, intensive gaping occurs.

As is shown by the first protocol, large objects can project some way above eye-level without eliciting gaping. With smaller objects, the boundary between ineffective and effective presentations is often exactly at the nestlings' eye-level.

Portielje (**5**) describes how young orioles (*Oriolus o. oriolus* L.) approximately 30 days old accepted food only when it was held above the beak. Food presented lower than this was fixated, but not eaten. Even when they had began to eat independently, they still paid attention only to fruit hanging above them. Since an adult male also behaved in this way, Portielje's observations do not apply only to the gaping response. The attachment to 'objects above eye-level' is, however, just as pronounced as in gaping blackbirds and thrushes.

SIZE

The nestlings gape towards objects of many different sizes. At first there seems to be no upper limit, for young nestlings respond strongly to a 20 × 40 cm rectangular, uniformly coloured board presented 20 cm away. But older nestlings no longer respond to very large objects. Thus, the upper limit changes during development. We believe that the possibility of a learning process can be excluded. Since we ourselves appeared in front of the birds before every feeding, any conditioning of the response would be expected to be to objects of our size, and the upper size limit should if anything be pushed upwards. It therefore seems that the change is due to maturation. This is not a maturation of the form of the motor pattern (to which the term 'maturation' is usually applied), but of the stimulus–response association in the releasing mechanism.

17.5.36 *Song-thrushes, 17 days old.* Vigorous gaping at white, and then at black circular cardboard discs of 20 cm diameter.

19.5.36 *The same birds, 19 days old.* No gaping response towards black discs of 18 cm diameter and of 16 cm diameter. In fact, the animals showed mild crouching.

As is indicated by the last protocol, the upper limit may alter through the fact that the association between certain visual stimuli and the response of crouching matures at this time, and (as in the brood that gaped initially in response to the alarm call) crouching begins to suppress gaping.

The minimum size can be determined quite well; it is about 3 mm. However, with such small objects the responses were fairly variable, and no firm conclusions can be drawn.

27.5.37 *Song-thrushes, 12 days old.* A green plasticine ball of 3·5 mm diameter suspended from a thread evokes gaping. A wire of 0·7 mm diameter does not release gaping; neither does white sewing yarn, though the birds do fixate it.

29.5.37 *The same birds, 14 days old.* No response to the white thread; weak gaping towards the 3·5 mm large plasticine ball. Vigorous gaping is elicited by a similar plasticine ball of 8 mm diameter.

Later the same day:

3·5 mm—weak response; 12 mm—vigorous response.

COLOUR[1]

In general, colour seems not to contribute to the releasing value of an object. We do not know the extent to which the birds are able to distinguish colours at an age when they still gape. We are not aware of any response of the two species investigated which is innately attached to a specific colour. It seems to be very difficult, or even impossible, to bring about any kind of colour-conditioning during the nestling phase. We attempted to condition our hand-raised birds to colours associated with ourselves. We always worked in white laboratory coats. Since the adults of both species are quite dark (and usually appear to nestlings in silhouette), we had a suitable opportunity to test whether our birds became conditioned to gape at white objects.

We first presented, either simultaneously or successively, a black object and a white object similar in every other respect. In these experiments black was always clearly preferred, not only by the blackbirds but also (though less markedly) by the song-thrushes. The

[1] *Author's note*: All this was based on pilot tests which we did not consider worth elaborating. In this section, 'colour' includes, and in fact is limited to, the brightness of our dummies.

latter is understandable when one considers that, in this case too, the feeding adult bird will usually be presented as a silhouette against the light surroundings.[1]

27.5.36 *Song-thrushes, 11–12 days old.* A white and a black cardboard disc of 8 cm diameter are presented side-by-side. Each dummy is presented three times on the right and three times on the left in order to exclude any left/right preference. The birds gape five times at the black disc and once at the white disc.[2]

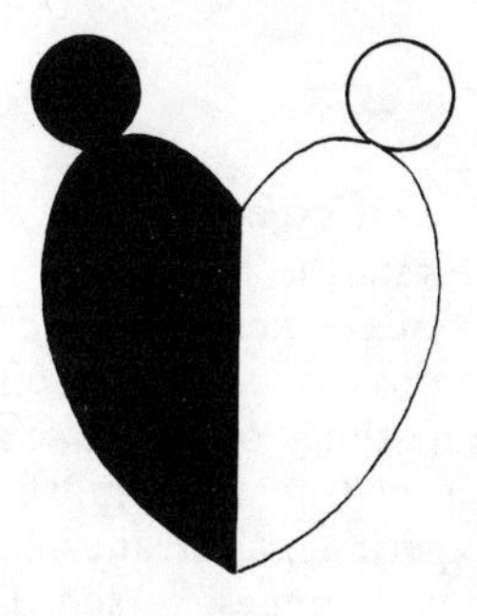

Fig. 116. Dummy used for investigating black/white preference.

Various broods of blackbirds were presented with the dummy illustrated in Fig. 116 and with a mirror image of this design. When care was taken to hold both 'heads' at exactly the same height (see later) the birds always preferred the black half.

SHAPE

As has already become clear, shape has no demonstrable influence on the intensity of the response. We found no shape which would not elicit gaping when moved above eye-level. There is no point in demonstrating this with protocols; the correctness of the assumption is immediately obvious from the protocols already provided.

This fact leads us to the supposition that free-living blackbirds and thrushes must possess the capacity for narrowing down the range of stimuli that release gaping through conditioning. If the

[1] *Author's note*: The higher eliciting power of black is so pronounced that we refrain from giving exact figures.

[2] To be exact, the experiment illustrated in Fig. 116 is concerned only with the orienting effect and not with the releasing value of the colours. However, for various reasons, we believe that it is best to describe this choice experiment at this particular point.

'innate releasing mechanism', with its limited character discrimination, were to remain completely unaltered, a mere breeze should continuously elicit responses from these forest-living birds, since there are usually waving leaves, branches and the like nearby. We did not study this question further, but it is quite certain that young, free-living blackbirds and thrushes are not stimulated to gape by the movements of nearby leaves. We do not know any more about this conditioning process.

MOVEMENT

We found that two different aspects of movement contribute to the elicitation of gaping. First, the adult bird must definitely exhibit movement; immobile objects never elicit gaping. Secondly, when a movement *across* the optical axis of the offspring is unsuccessful, movement *towards* the nestlings often evoke a response. In order to investigate the influence of transverse movements, we initially employed the following experimental arrangement.

We placed one of our dummies, attached to a stand, above the brood under investigation. The lid of the box was then carefully removed, and we observed whether the immobile disc evoked a response. If it did not, we moved the disc and counted any responses shown after the movement as positive. Only in a very few cases did we observe responses with the immobile disc (as we discovered later, these were released not by the disc itself but by the movement

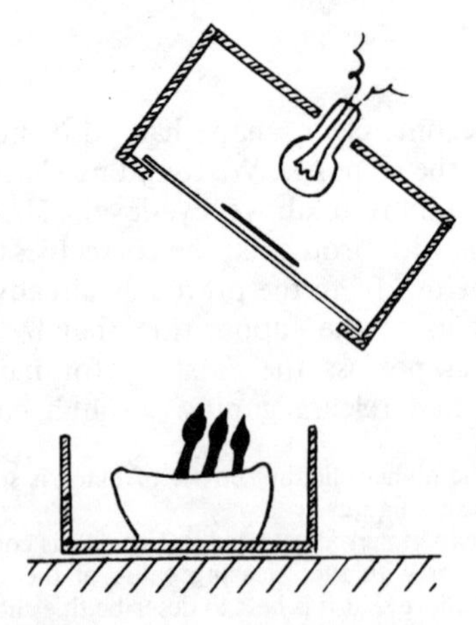

Fig. 117. Arrangement for projection of silhouettes.

of the lid, since sometimes lifting of the lid evoked gaping even in the absence of a dummy).

We then used silhouettes (Fig. 117) which could be presented at any chosen moment by switching on an electric light behind a model held by very thin wire, which in turn was screened from the birds by an opaque 'through-projection screen' of ground glass. Thus a silhouette could be produced without any movement visible to the birds. With such presentations, immobile silhouettes never elicited a response, but even slight movement did evoke gaping.

26.5.36 *Song-thrushes, 10–11 days old.* Light behind model extinguished. Lid of box removed, followed by pause until birds have settled. Light then switched on and off four times. At each change, the birds look upward somewhat, but do not gape. The disc is then moved gently to-and-fro behind the opaque glass; gaping immediately follows.

The response-eliciting movement *towards* the birds, which is considerably more effective, can best be demonstrated when the birds have been so well fed, or the gaping response has waned so much, that no more responses are elicited by a movement perpendicular to the optical axis. If the dummy is then moved towards the nestlings, they will often gape again. This is probably not an effect of increasing size of the retinal image since the difference remains when, in a control experiment, the transverse movement and therefore the change in the retinal image is made much more extensive than the movement caused by approach. Also, as we have seen, differences in size of the dummies had little, if any effect. Thus the birds definitely showed depth-perception. The approach of the dummy is most effective when it comes obliquely from above.

4. Orientation of the Response

Gaping is always oriented with respect to the surroundings. Such orientation does not remain constant throughout the whole nestling-phase; two successive phases can be distinguished. In the first period, the neck is stretched vertically; in the second period, the neck and gape are directed towards the adult bird's head. Strangely, the transition from the first to the second period does not coincide with the onset of visual release of the response. It is only one or two days after gaping can first be elicited visually that visually-determined orientation appears. Thus, if we use both aspects (release and orientation) we must distinguish three phases: gravity orientation with mechanical elicitation; gravity orientation with visual elicitation; visual orientation with visual elicitation. Before considering this fact, which would appear to be theoretically important, the three periods must be more exactly defined.

GRAVITY ORIENTATION WITH MECHANICAL ELICITATION

As noted above, this period lasts 9–10 days. The fact that orientation is controlled by gravity is inferred from the following evidence: we found no changes in light conditions which could influence the direction of neck-stretching. Neither the direction of incident light, nor the movement of light or dark objects, nor even exposure to total darkness have any effect whatsoever. Other mechanical stimuli, such as tilting the nest or mechanically obstructing the normal movement, have no effect either. For example, the birds repeatedly struck the lid of the box with their beaks when we tapped the closed container, and did not correct their movements.

GRAVITY ORIENTATION WITH VISUAL ELICITATION

The length of this period varies greatly between individuals. It usually lasts only 1 day, but sometimes lasts up to 3 days. On average, this period commences on the 9th day in the blackbird and on the 10th day in the song-thrush.

At this time, the birds still frequently lie in the nest with their eyes closed. As has been stated, they can then be induced to gape only by shaking or touching. However, once they have opened their eyes in response to jarring, visual stimuli can indeed release gaping, but even then they do not have the slightest influence on the direction—the birds gape vertically upwards, frequently bypassing the eliciting object. An extreme case can be provided as an example.

24.5.37 *Song-thrushes, 9 days old.* The eyes are closed, visual stimuli have no effect; touching the nest-rim evokes a vertical gape. Later on the same day: I move my finger slightly above eye-level. One bird is lying with its eyes open and immediately extends its head to gape beyond my finger. The others are stimulated by this movement and similarly gape vertically upwards.

In the course of the next two days, the birds orient more and more to visual stimuli. The first indication of visual orientation is observed on the 12th day.

VISUAL ORIENTATION WITH VISUAL ELICITATION

As previously stated, the young birds begin to direct their gaping movements towards the adult's head 1–3 days after their eyes have opened. That this orientation is indeed entirely dependent upon visual stimuli, will be shown below.

The change to visual orientation to the head of the parent takes place gradually. Initially, the nestlings still gape almost vertically, but point slightly towards the adult. In the course of 12–24 hours (i.e. quite gradually), this switch is completed, and the neck of the nestling is directed precisely towards the adult's head. Thus, the resultant orientation of gaping is due to the gradually changing relative importance of mechanical and visual cues.

Preliminary experiments conducted in 1935, which had already given suggestive results, were later followed by the experiments described below. In order to evaluate the orienting effect of various cues we either made the nestlings choose between two simultaneously presented dummies differing only in one characteristic, or one part of the dummy was moved in order to see whether the birds would follow the movement.

RELATIVE HEIGHT

If the birds are presented with a circular cardboard disc whose diameter is not too small (e.g. 8 cm), their gaping is not aimed at the centre, but at the upper margin. In this and in the following experiment, we always ensured that the entire disc was above eye-level.

Fig. 118. Presentation of two rods at different heights, but at the same distance.

Similar experiments were done with two rods (Fig. 118) and with two fingers. In order to control for any other differences, we presented alternately each of the two objects as the higher. In every case the birds gaped exclusively at the higher dummy. The higher object did not have to be completely above the other; it was sufficient that only part of it was higher than the highest point of the other object, and a difference of only a few millimetres was quite adequate. From now on, we shall refer to the 'higher dummy' as that whose highest parts protrude so far above the other dummy that they act as a focus for gaping.

16.5.36 *Song-thrushes, 16 days old.* Two black rods at the same distance, one above the other. Gaping directed towards the higher (Fig. 119).

Fig. 119. Black double disc.

19.5.36 *Song-thrushes, 19 days.* Black double disc (Fig. 119). As soon as, by tilting the dummy, we raised the upper edge of one half to 3 mm above that of the other, the birds aim at the higher rim.

19.6.36 *Blackbirds, 16 days old.* Two rods presented at the same distance are held so that one is 3 mm higher than the other. The birds gape towards the upper rod.

RELATIVE DISTANCE

If two visually similar objects are presented at the same height, the birds will gape towards the nearer of the two (Fig. 120). This effect, too, can be obtained with a wide variety of dummies. However, since the birds respond so sensitively to height, one must ensure that the two dummies are presented at exactly the same level.

16.5.36 *Song-thrushes, 16 days old.* Two visually similar rods are presented at the same height and the same distance. With repeated presentations, the birds gape sometimes towards the left rod and sometimes towards the right, but if one rod is held at a distance of 3 cm, and the other at 2·5 cm, gaping is immediately directed towards the latter.

We also frequently observed the following

30.5.36 *Blackbirds, 10 days old.* The birds gape at one another's heads as soon as the dummy is farther away than the head of a sibling. Frequently, the gaping response is initially released and oriented by the dummy, but then becomes directed towards the siblings. We attribute this to the fact that the dummy is held relatively still, while the heads of the siblings are moving vigorously. Often, the central bird gapes towards the dummy, whilst the two birds on either side gape towards the former's head. The fact that the middle bird

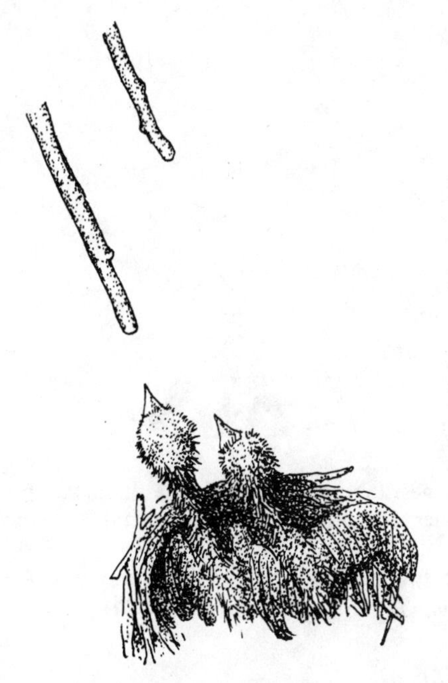

Fig. 120. Gaping directed at the nearest of two rods at the same height.

does not gape towards his neighbours is a result of our method of presentation. The animals at the side must always gape obliquely, so that their heads are automatically lower than the head of the bird at the centre (Fig. 121).

If distance and height are varied independently, the effectiveness of height is once more demonstrated:

14.5.36 *Song-thrushes, 14 days old.* When two black rods are presented at the same height, the nearest is always chosen. One rod is then presented at a distance of 3 cm, whilst the second rod, which is 3 cm higher, is held 2·5 cm further away. Nevertheless, the higher rod is still preferred. The lower rod is then presented at a distance of

Fig. 121. One nestling gapes towards the dummy, while the birds on either side gape towards the central bird's head.

4 cm, with the higher rod 4 cm further away and 4 cm higher. The birds gape towards the higher rod (Fig. 122).

RELATIVE SIZE

If two cardboard discs of different size are presented at the same height, the birds direct gaping towards the smaller one. If the small

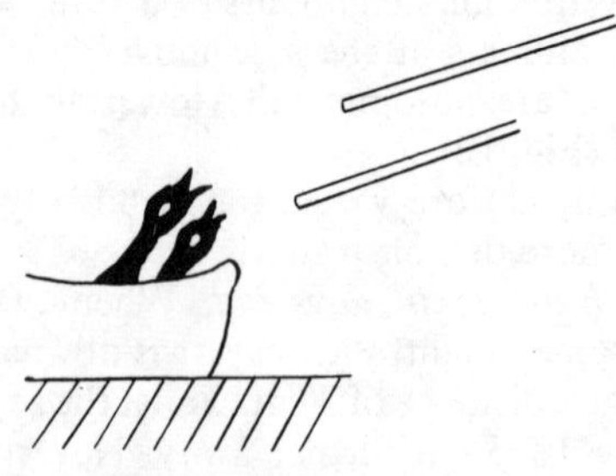

Fig. 122. Two rods. Height in competition with distance.

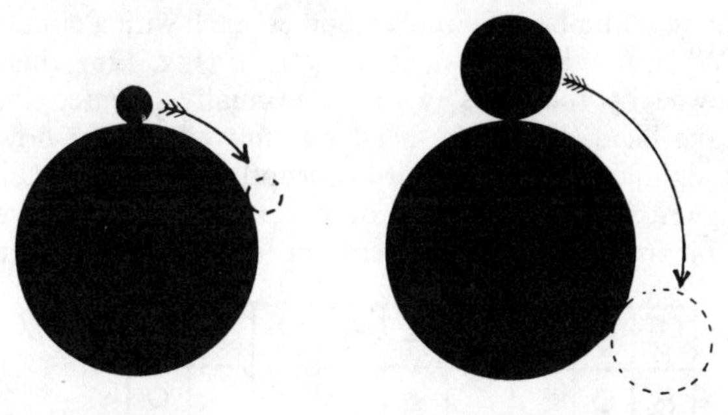

Fig. 123. Maximum permissible rotation for two heads of different sizes.

disc is considerably smaller than the other, gaping is directed towards the small disc even when it is situated below the larger. Thus it would seem that the size of the 'head' *in relation to* that of the 'body' is decisive, not its absolute size. In order to study this in more detail

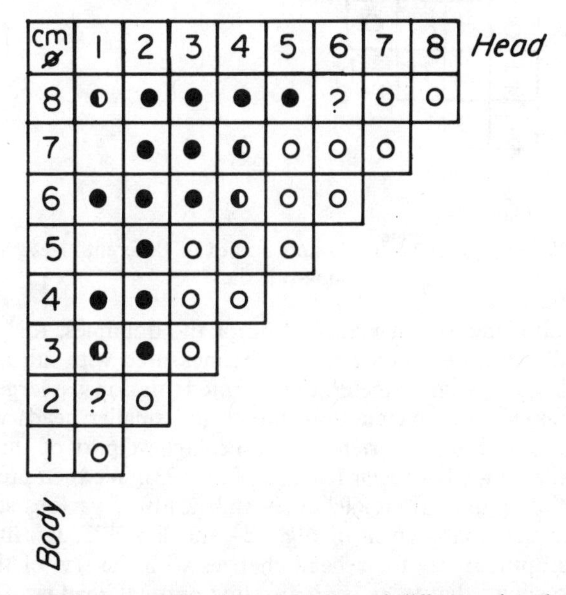

Fig. 124. Responses to heads and bodies of different sizes by young blackbirds. White circles: the higher is always selected. Half-black circles: the smaller is preferred, even when 'lower' than the larger. Black circles: the smaller is markedly preferred and is followed some distance laterally. ? = indecisive result.

we presented a number of circular 'bodies' each with a circular 'head' on top. When the dummy was then rotated (Fig. 123), this rotation was followed by the birds, which continually oriented themselves towards the head. But if the head was moved too far downwards, it lost its significance and the birds abruptly re-oriented their gaping to the highest part of the 'body' outline. We measured the extent of 'permissible rotation' for different 'head' and 'body' sizes. With

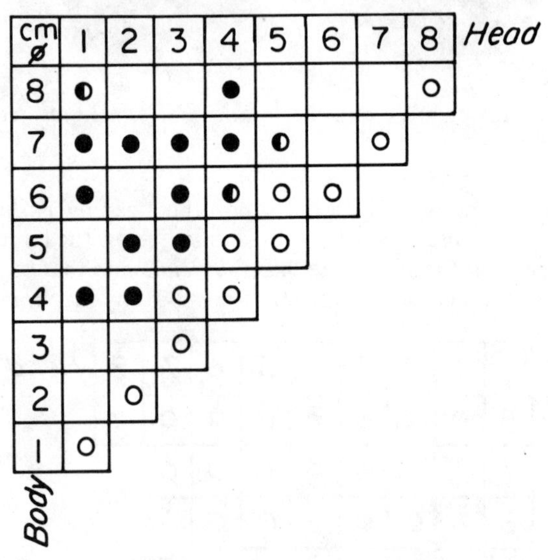

Fig. 125. Responses to heads and bodies of different sizes by young song-thrushes.

fairly well standardised presentation of the dummies, this angle of maximally tolerated rotation could be measured with fair accuracy. With a body of 8 cm diameter, the tolerated rotation was largest with a head of 3 cm in diameter. Both larger and smaller heads were disregarded (the birds re-orienting to the highest part of the 'body') before this angle had been reached (Fig. 123). We then presented a number of combinations of bodies and heads of various sizes, and obtained the results given in Fig. 124 and Fig. 125. It can be seen that the optimal size for a head changes with the size of the body. This too shows that there is no absolute optimal head size, but only one relative to body size. The optimal head size seems to be roughly one-third of the diameter of the body.

The same was then demonstrated in still another way. We presented a body with two heads of different sizes, the upper margins of

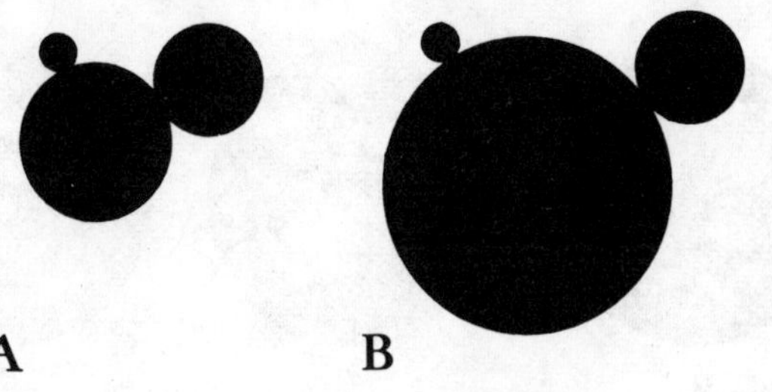

Fig. 126. Two dummies with double heads. Body of (A) 4 cm; (B) 8 cm in diameter. Heads in both dummies: 1 cm and 3 cm in diameter.

which were at exactly the same level. The birds then gaped at one of the two heads. With the two head-sizes used (1 and 3 cm in diameter) the young birds preferred a head of 3 cm diameter on a body of 8 cm, and a head of 1 cm on a body of 4 cm diameter (Fig. 126).

A further question is: how small can the head be before the birds cease to aim at it? Since, as will be shown below, the shape of the head is of minor importance, the series of dummies used could be given a wide variety of heads.

First, we gradually 'eclipsed' a head of 1 cm diameter behind a body of 8 cm diameter, in order to see how far the head must protrude in order to have an orienting effect at all. The crucial point was reached when the head protruded only 2 mm (Fig. 127).

A second series of dummies was constructed by drawing two tangents to the body circle from a point which was placed increasingly close to the body, such that the angle enclosed by the tangents became increasingly large. The smallest interruption of the smooth outline which still attracted the birds' attention is shown in Fig. 128*a*.

With this technique, we demonstrated a gradual change with age. The older the animals were, the slighter was the protrusion to which they would respond. We did not determine whether a learning process or maturation was involved.

Fig. 127. The smallest protuberance to be still effective: 2 mm of the head circle still protruding.

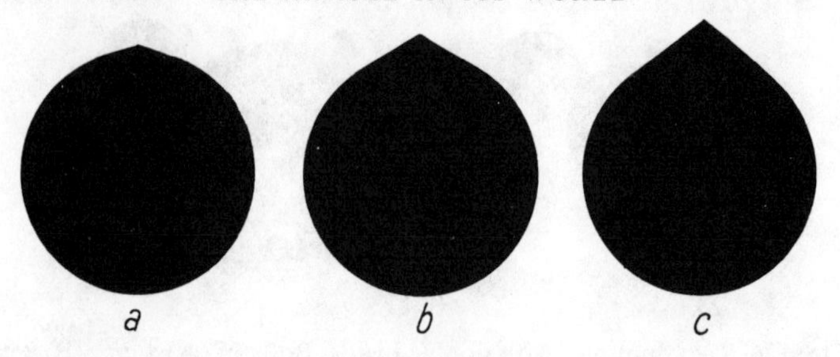

Fig. 128. Three head dummies; left: the smallest protuberance to which the birds responded.

SHAPE

Experiments with fingers, rods, etc. had already shown that the shape of the 'head' can vary greatly without losing its power as an orienting stimulus. In order to obtain more precise information, we presented a number of identical circular 'bodies' with heads of various shapes (Fig. 129). All the heads had approximately the same surface area as

Fig. 129. Five different heads of equal attractiveness.

the circular head which was optimal in size with respect to the body circle employed.[1]

All of the heads illustrated here, as far as our crude methods could demonstrate, had the same effect, releasing as well as orienting. When we rotated the body, no difference in the degree of permissible rotation was seen, nor did we see signs of any change with increasing age; but we did not carry out choice experiments, which might have revealed slight changes.

Thus, according to these experiments, the nestlings appeared to regard all discontinuities in the body outline as heads. We next wondered whether such discontinuities had to be protrusions or

Fig. 130. Concave interruption of outline.

whether concave interruptions, or indentations would also work, and we therefore presented the model illustrated in Fig. 130, which has a 'dent' rather than a head.

To our surprise the birds gaped at the dent as well as at a 'head'. Later we discovered that the orientation to this 'concave outline discontinuity' appeared approximately one day later than that to the optimal convex heads. But on closer observation of the gaping movements we also noticed that it was not really the dent, but the areas of the 'body' left and right of it to which the birds pointed. If the model was rotated so that the indentation passed from left to right of the 'zenith' (or vice versa) the birds abruptly and markedly switched from one side of the dent to the other as the zenith was passed, each time gaping towards the higher of the two. Thus to the birds the model would seem to represent a 'body with two heads', i.e. two protrusions. This explanation would be completely satisfactory if it could be shown more clearly that convex discontinuities are more effective

[1] Since we did not know which characters are utilised by the birds to determine the size of a 'head' (surface area, total outline, etc.), and since the above-mentioned experiments on relative head-size had shown that optimal size is not sharply demarcated, it would have been arbitrary and quasi-objective to standardise 'size' according to any more precise criterion.

than concave ones of the same dimensions. This actually proved to be the case with the models represented in Fig. 131.

With the model *a*, the birds always gaped towards the protruding head. If this head was lowered by rotating the model, the birds switched to the 'bulge' above the notch. In the model *b*, the notch has two advantages over the protrusion: first it is higher, and secondly it departs somewhat more from the outline. Nevertheless,

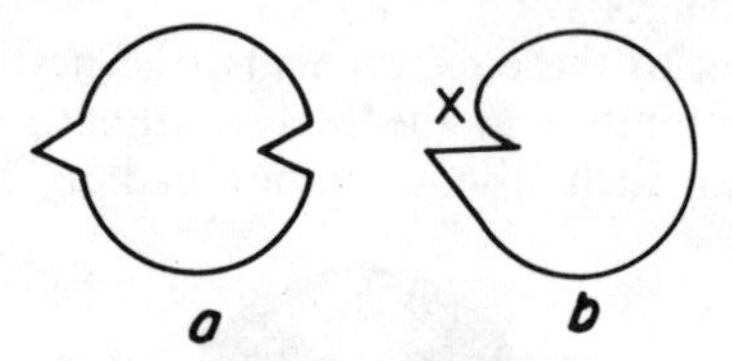

Fig. 131. Concave and convex discontinuities are presented together.

Fig. 132. Two-headed dummy; one head with neck-notch, the other without.

the birds gaped towards the spike (convexity). If the dummy was rotated further downwards, gaping switched to the point marked X.

Thus, convex disruption of the outline, or 'bulge' is a characteristic of a 'head'.

However, the outline of the adult bird does have a significant concave discontinuity—namely that of the boundary between head and body ('neck'-notch). When the nestlings were forced to choose between two heads, one with and one without a neck-notch, they always gape towards the head bounded by a notch. The smallest notch effective with a body disc of 8 cm is one 2 mm deep (Fig. 132).

BRIGHTNESS CONTRAST

When we presented a black circular disc with a small white spot near the margin (Fig. 133, left), the birds gaped towards the spot.

However, a black spot on a white circle was equally effective (Fig. 133, right). This was true of both species. When the discs were rotated, the young birds followed the spot as if it were a bulge.

14.5.36 *Song-thrushes, 14 days old.* When the discs (shown in Fig. 133) are rotated the nestlings follow both spots down through approximately 90°, but no further.

17.5.36 *Song-thrushes, 17 days old.* Result as on the 14th, except that one young bird repeatedly follows the black spot on the white disc through an angle of 180° (i.e. down to the lowest possible position).

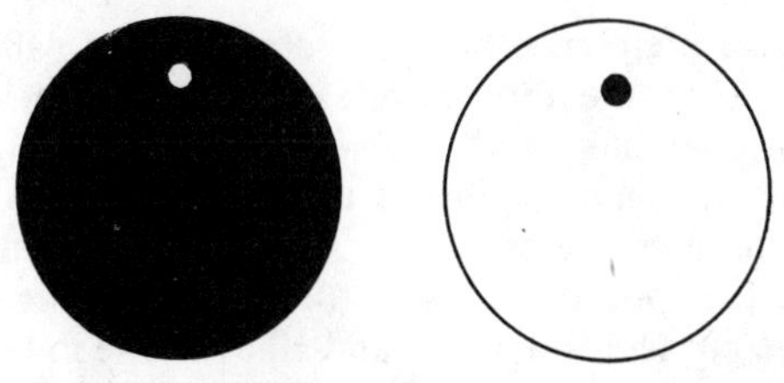

Fig. 133. The two dummies with a contrasting spot.

30.5.36 *Song-thrushes, 14–15 days old.* White and black discs 8 cm in diameter with black and white spots respectively (1 or 2·5 cm) are rotated. The birds follow the spot almost without exception the whole way down. However, when such a model is presented with the spot already in the lowest position, the nestlings gap towards the upper margin of the disc.

Since in both species the head colour of the parent bird is scarcely different from the body coloration, it is difficult to imagine how this observed effect of the coloured spot fits in with the natural situation. It might conceivably help them to aim at the bill, which, at least in the male blackbird, stands out by its distinct orange-yellow colour. If we were dealing with a specific response to the bill (or perhaps to the food), the optimal size for a spot should be smaller than that for a 'head'. Therefore, in the following experiments we presented two spots of different sizes on one body (Fig. 134). We were unable to demonstrate a smaller optimal size for contrasting spots—to the contrary gaping was always directed at the larger spot.

23.6.36 *Blackbirds, 14 days old.* Two white spots of 0·8 and 2 cm diameter are painted on a black circular body of 8 cm diameter. The birds always select the larger spot, even when it is presented somewhat lower than the small spot.

29.7.37 *Blackbirds, 12 days old.* The same experiment—black spots on a white disc. Same result.

Fig. 134. Large and small contrasting spots at the same distance from the disc margin.

Thus, the size preference bore no relationship to the real size of either the beak or the food, while it does correspond, however crudely, to the parent's head. However, we had a slight suspicion that the birds were not gaping at the spot itself, but at its upper margin. It sometimes seemed that gaping was directed at the black 'horns' above the spot, just as with the concave outline discontinuities, where the corners to the left and right of the indentation acted as 'heads'. As can be seen from Fig. 134, for the human eye at least, the large spot produces far more conspicuous heads than the small spot, at least as long as both spots are equally near the edge. If this were how the birds perceived it, it was to be expected that the further a spot was from the outer margin, the less influence it should have, since the 'horn' effect is greatest when the spot is closest to the external margin.

We therefore presented the model shown in Fig. 135, and here the smaller spot did in fact prove to be more effective. This result was obtained with both black and white discs. In the experiment with the white disc, the blackbirds differed slightly from the song-thrushes in their response. The blackbirds often gaped towards the smaller spot when it was some distance from the outer margin. This is possibly to be explained from the greater responsiveness to black as such, exhibited by the blackbirds from the very beginning of the experiments. The small black spot might possibly have acted as the 'head'

Fig. 135. The large spot is farther from the edge than the smaller spot.

of the adjacent larger spot, if the white disc were ignored. In order to test this possibility, we presented, instead of two black spots on a white disc, two black circular discs of exactly the same size and relative distance from each other as the spots had been (Fig. 136).

If the white disc in the previous experiment had merely been acting as background, the experiment with the two black discs should give exactly the same result as that with the black spots. The result was inconclusive. Sometimes the birds gaped at the small disc, sometimes at the larger. So we cannot explain why blackbirds sometimes chose the small black spot. Perhaps the two discs are sometimes responded to as two separate objects (i.e. as two adult birds of different size), and sometimes as a single 'body-with-head'.

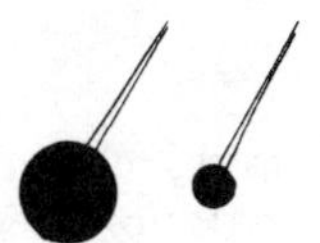

Fig. 136. Presentation of two 'spots in isolation'. The discs were attached to a crossbar (not illustrated) in order to keep the distance between them constant.

With the exception of the occasional aberrant choices of the blackbirds, the reported results permit us to conclude that the contrasting spots are effective because they provide—in combination with the nearby outer margin of the 'body'—two adjacent 'heads', each acting as a bulge. There would therefore seem to be no reason to regard the responses to contrasting spots as different from those to disruptions of an outline. The objects orienting the gaping (the 'heads') acted as such by virtue of their contrast with the background, and further by being bulges of a certain relative size and level, while their shape was relatively unimportant.

5. Control Experiments in the Field

To meet a number of possible objections, our experiments had to be checked with wild birds raised by their own parents. In the first place, deficient feeding of our hand-raised birds might have lowered the threshold of the gaping response, and this could have resulted in the observed lack of specificity. Secondly, the hand-raised nestlings could have become conditioned to us.

We therefore repeated the critical experiments with a fair number of wild birds (16 blackbirds and 10 song-thrushes), which were reared entirely by their own parents and were only occasionally shown models in the absence of their parents. We never removed these birds from their nests, and always ensured that the readiness to gape was not inhibited by parental alarm calls during our experiments. The following experiments were repeated in the field.

A. ELICITATION

Impact on the nest-rim; touching of the beak margin; sound; visual stimuli (fingers; rods; black, white and brown cardboard discs of various shapes and sizes; choice between black and white discs; relative height).

B. ORIENTATION

Differentiation of the three phases; relative height; relative distance; relative size (choice experiments); non-specificity of head shape; the head as convex interruption of the outline; neck-notch; choice experiments with contrasting spots of different size and location.

The result of these field controls can be summarised rapidly: all experiments had almost exactly the same results as those in the laboratory. We found only the following (in our opinion negligible) differences. As already mentioned above, one brood responded to the father's call. In addition, the naturally reared young birds usually appeared to be somewhat less hungry than our nestlings; they did not gape so vigorously. However, this had so little effect on the elicitation threshold that with our methods no difference in response readiness could be determined.

Thus, the field experiments dispose of the objections regarding abnormal conditions, either with respect to threshold-lowering or damage due to inadequate nutrition. Nor did they reveal any evidence of conditioning of the gaping response.

The question whether the gaping response is normally altered by learning processes at all is not quite settled by these experiments; it will be discussed in the last section.

6. Evaluation and Summary

Our original intention was to conduct a thorough investigation of an 'innate releasing mechanism'. In doing this, the aim was not limited to demonstrating the existence of simple 'schemata' (von Uexküll–Lorenz) with only very few unspecific 'sign stimuli'; we also wished to analyse these stimuli as far as possible. However, it soon emerged that there is no such thing as a 'schema of the parental companion' ('Schema des Elternkumpans' Lorenz (**2**)) and in fact not even a 'schema of the food-donor'. For this single response of gaping there are two collateral schemata, one orienting and one merely releasing. In both cases, maturation occurs in both the motor and the receptor components.

The stimuli releasing the gaping are initially mechanical in nature; later, visual factors are added. For visual elicitation, an object must be larger than 3 mm across; it must be above eye-level, and it must preferably move towards the nestling. The form of the motor pattern also changes: the extension of the neck becomes more pronounced, and—as described—wing and leg movements are added.

The stimulus orienting the response is similarly mechanical at the beginning; the nestling orients itself to gravity. This mode of orientation later gives way to visual orientation; gaping is then directed towards the adult bird's head. A 'head' is characterised by movement, contrast, height, size in relation to body, nearness; and it has to be a convex interruption of the 'body's' outline, preferably separated from the body by a 'notch'.

The complex character of the gaping response persists throughout the entire nestling period; there are always two mechanisms operating together. We are therefore forced to make the same distinction as that made by Lorenz (**3**) between 'instinctive motor pattern' or 'fixed action pattern' ('Erbkoordination') and 'orienting response' ('Taxis'). As in the egg-rolling response of the Greylag goose analysed by Lorenz and Tinbergen (**4**), the two mechanisms act simultaneously. An important methodological point in the present study is that we have distinguished between fixed action pattern and taxis in a manner different from that employed with the egg-rolling response of the Greylag goose. In the first place, our findings showed that the transition from mechanical to visual stimulus situations occurred earlier in the releasing than in the orienting mechanism. Secondly, the optimal stimulus situations of the two mechanisms are different, both in the mechanical and in the optical fields.

It is evident from what has been said that a 'schema' can just as well be associated with a taxis as with a fixed action pattern. This leads us to remark upon a certain misuse of the word 'schema', which is based upon a certain vagueness in conceptualisation. References to 'the schema of the sexual partner', 'the schema of the parental companion' and so on do not always mean the same thing. Sometimes they apply to the mechanism controlling *one* specific response directed towards the sexual partner, parent and so on; but on other occasions they are applied with reference to the sum total of *all* responses directed towards the sexual partner, etc. However, since the behaviour repertoire of an animal often incorporates several responses directed towards the sexual partner, etc., and since each of these responses may be, and usually is, controlled by different stimuli, the words 'releasing mechanism' and 'orienting mechanism' should be applied to *responses* rather than to outside *objects.* The alarm call of the parent elicits a different response in the young from the arrival with food, yet both convey messages from the same outside object, *viz.* the parent. Secondly behaviour patterns are often linked in succession, forming a chain of components, each of which is controlled by different stimuli. Examples are provided by the mating of many animals (birds, fish, insects). Proof of this chain structure is provided by the fact that, if certain stimuli are not forthcoming, the behaviour is always arrested at quite specific points. The gaping response itself is the first component of a chain consisting of at least two links: the next component is swallowing, which only occurs when there is actually something in the throat. The tactile and taste stimuli which are concerned are just as much a part of 'the schema of the parent' as the stimuli releasing the gaping response which have been analysed above.

Thirdly, our experiments show that, apart from successive linking, a releasing and an orienting mechanism can control what looks like one single response, but turns out to be two simultaneously performed activities.

It must be emphasised that the question of the possible alteration of the mechanism through learning is irrelevant with respect to this conceptual distinction of the two components of gaping. Nevertheless, it is intrinsically important to ask whether we were investigating innate shemata, and this therefore merits brief discussion. There are in fact very few analyses that show such schemata to be innate.

Since our laboratory experiments had the same results as the field tests, the argument that our experiments must have involved conditioning to features distinguishing us from the normal food-donors can be discounted. However, in many other characters there was a general similarity between us and the adult birds, so that conditioning to some cues could have occurred equally in the field and the laboratory, and our field checks could not have demonstrated the effect. However, the laboratory experiments themselves did give us some information, which, when combined with the experimental data from the field, permit a reliable judgement on possible learning processes.

The mechanical stimuli which release gaping are effective even when the birds have had no prior experience. In addition, it is impossible to condition the animals *not* to gape in response to jarring the nest. Likewise, the responses to visual releasing stimuli, such as size, and blackness are definitely not learnt: specific sensitivity to these cues was evident the moment the nestlings opened their eyes and therfore before they had had any opportinity to learn visual patterns. As described, the response to size alters during development, and it has already been demonstrated that no learning process is involved here.

Since both the adult birds and the human investigators always gave the nestlings food from above, there is of course the possibility that the response to the cue 'above eye-level' may have been learned. We therefore tried to teach the nestlings to gape also towards objects which we moved around below eye-level. This was done by inducing gaping through jarring the nest and then placing food in the birds' beaks from below. However, such birds did not learn to gape towards objects moved below eye-level. The preference for black exhibited by our experimental animals also speaks against learning: we always fed the birds in white laboratory coats, and yet the nestlings did not develop a preference for white.

As already mentioned above, we believe that learning processes do occur which render certain stimulus combinations *in*effective (i.e. 'carving out' so-to-speak a small area from the previously broad innate schema). For example, moving twigs and leaves—which possess all the visual characteristics of the optimal stimulus situation —do not evoke gaping, and this can only have been brought about by experience. In order to observe this process in the field, it would have been necessary to conduct continuous observation of a number of broods; unfortunately we omitted to do this.

From our experience, we doubt whether positive conditioning of the gaping response is at all possible, for example teaching the birds to gape more intensively towards forceps than to the dummies. It is emphasised that all of these conclusions are relevant only to the two species investigated, and are restricted to this one response. It is, of course, to be expected that there will be differences between species and between different responses of one and the same species.

Of course, the young of many other song-birds do learn to gape more vigorously towards certain stimulus combinations than is 'prescribed' by the innate mechanism. For example, various species doubtlessly learn to recognise their individual parents or foster-parents (cf. Lorenz (**2**)). However, this learning process does not presumably affect gaping as long as the young are in the nest; if such learning processes do occur with the two species we investigated, they must occur after fledging.

Finally, in many cases a completely different type of learning process can give the impression that there has been an alteration in the releasing schema of the gaping pattern. Freshly captured young animals often crouch in response to human presence. When the animals then become accustomed to the new situation, the 'block' on gaping is removed. It could easily be inferred that the animals have 'learned to gape towards the forceps'. However, it seems far from unlikely that with many passerines the pair of forceps fits excellently into the innate schema of the gaping response. But in this case, one cannot speak of a learning process affecting the schema of this response; rather it is habituation, of the crouching or the escape response, which is primary, and this in its turn lowers the inhibiting effect on gaping.

As far as orientation is concerned, the reaction to gravity is of course innate, since it is exhibited even by newly hatched nestlings. In the visual modality, the cue of relative height belongs to the innate schema, since even our animals—which were not fed from the highest part of our bodies—developed this preference for the highest point. We cannot say anything about relative distance, since both the control animals in the field and the laboratory animals always received food from the part of the food-donor nearest them. In both cases, they could have become conditioned to this cue. However, the response to 'relative size' must be innate. Our experimental animals could never have learnt this from us, since our hand does not bear a 1:3 relationship to our body. Shape doubtlessly plays a very small part (or none at all) in the innate schema. We did not investigate how far conditioning to certain shapes is possible, but if such conditioning were usual in the development, we ought to have found changes in responsiveness to the differently shaped dummies we used—in favour of the food-providing hand-and-forceps and to the detriment of the many oddly-shaped dummies which did not provide food.

From these considerations, it would definitely seem that the optimal stimulus situations which we determined represent largely an 'innate releasing mechanism' in Lorenz's sense, one which has not been produced to any demonstrable extent by conditioning, and which in several respects has even withstood conditioning attempts with remarkable persistence.

Probably the most extensive study published so far on the elicitation of gaping is that of Kuhlmann (**1**). This author describes for different passerines (*Agelaius p. phoenicus* L. and *Hylocichla mustelina* G. M.) alterations in responsiveness, which he in one case ascribes to learning and in another to maturation, though without giving reasons for the interpretation in each instance.

To our knowledge, there is only one other case in the available literature where both the releasing and orienting stimulus situations of a response directed at the head of another animal have been investigated. Portielje (**6**) found that when a bittern is prevented from fleeing by a surprise approach, it will adopt a defensive posture and peck with lightning speed at the enemy's head as soon as the latter approaches too closely. However, when the enemy has no 'head', the jabbing movement is not performed: when Portielje tucked his head far down between his shoulders and covered himself with a cloth, the jabbing movement failed to appear. But as soon as he placed a crudely-head-shaped cardboard disc above the covering cloth, the bittern jabbed at the cardboard disc. From his experiments, he concluded: '*Botaurus stellaris*, when in the fright posture opposite an opponent *in toto* and prepared to defend itself, responds to a complex of characters, which probably consist of something like a small head shape or contour above a larger torso shape or contour' (p. 11). Portielje's experiments do not seem to demonstrate conclusively that the head must be smaller than the body, and located above it. But his experiments do show that the head shape itself needs to possess only a small number of characters, and, further, that a head is necessary not only for orientation but also for elicitation of the jabbing-movement. In this respect, the response of the bittern is different from the gaping response of blackbirds and thrushes.

REFERENCES

1 KUHLMANN, F. (1909). 'Some preliminary observations on the development of instincts and habits in young birds', *Psychol. Rev. Monogr. Ser.* **11,** 49–85.

2[1] LORENZ, K. (1935). 'Der Kumpan in der Umwelt des Vogels', *J. Orn. Lpz.*, **83,** 137–213, 289–413.

3[1] —— (1937). 'Uber den Begriff der Instinkthandlung', *Folia biotheoretica*, **2,** 17–50.

4[1] —— and N. TINBERGEN (1938). 'Taxis und Instinkthandlung in der Eirollbewegung der Graugans', I. *Z. Tierpsychol.*, **2,** 1–29.

5 PORTIELJE, A. F. J. (1922). 'Eenige merkwaardige instincten en gewoontevormingen bij vogels', *Ardea*, **11,** 23–39.

6 —— (1926) 'Zur Ethologie bzw. Psychologie von *Botaurus stellaris* (L.)', *Ardea*, **15,** 1–15.

[1] All three papers available in English translation in: *Animal and Human Behaviour*, Vol. I, Methuen (London, 1970)—a collected edition of some of Konrad Lorenz's earlier papers.

(*From the Zoological Laboratory of Leiden University and the Department of Zoology of Oxford University*)

12

The Spines of Sticklebacks (*Gasterosteus* and *Pygosteus*) as means of Defence against Predators (*Perca* and *Esox*) (1956)

1. Introduction

Many species of fish have evolved sharp spines which are often alleged to serve as defensive weapons against predators. Most frequently these spines are modified fin-rays. The sticklebacks possess this type, and it was the purpose of this study to analyse the effect of these weapons upon two predators, the perch (*Perca fluviatilis* L.) and the pike (*Esox lucius* L.). The species of stickleback studied were the Three-spined stickleback (*Gasterosteus aculeatus* L.) and the Ten-spined stickleback (*Pygosteus pungitius* L.), the only two members of the family which occur in European freshwaters. The reactions of the predators to these fish were compared with those to various non-spined freshwater fish, viz., the minnow (*Phoxinus phoxinus* (L.)), the roach (*Rutilus rutilus* (L.)), the rudd (*Scardinius erythrophthalmus* (L.)), and the Crucian carp (*Carassius carassius* (L.)).[1]

These two predators and five prey species are all known to occur together in the wild, and Hartley (**6**) has found sticklebacks, minnows and roach amongst the stomach contents of the predators.

Some preliminary tests by Tinbergen in 1939 appeared to show that young pike of approximately 15 cm long had considerable difficulty in swallowing adult Three-spined sticklebacks, and also that they soon learned to leave sticklebacks alone, while continuing

[1] Apart from roach and rudd we may occasionally have used other, similar species. Since we used young fish of approximately stickleback size which are not easily determined, and since the responses of the predators to all these non-spined fish were always the same, we did not attempt to determine the exact species. The majority of these fish were taken from waters in which larger roach and rudd were abundant.

to eat non-spined fish of the same and even considerably larger size. Since sticklebacks, when seized by a pike, respond by raising all their spines, and keeping them rigidly erected as long as they are in the pike's mouth, it was naturally supposed that the conditioning process in the pike had something to do with the presence of the spines. This problem was tackled experimentally by Hoogland in 1948. Later, in 1952, Morris, engaged in a study of the behaviour of *Pygosteus*, extended Hoogland's work studying perch as well as pike, and *Pygosteus* as well as *Gasterosteus*. The observations done so far throw light not only on the function of the stickleback's spines but enabled us at the same time to analyse the predator's behaviour to a certain extent. Also, striking differences were found between the two species of sticklebacks, and these could be correlated with differences in their reproductive behaviour and other aspects of their ecology generally.

A film of the predatory behaviour of pike towards sticklebacks and non-spined fish was taken by Tinbergen, and copies are in the Department of Zoology at Leiden and in the Department of Zoology and Comparative Anatomy at Oxford.

2. Material and Methods

Five pike and several hundreds of *Gasterosteus*, Rudd, roach and Crucian carp were collected near Leiden in Holland; three pike and five perch, as well as all sticklebacks and minnows used in Morris's experiments were collected from ponds and streams in Wiltshire and Oxfordshire in England. The prey specimens offered were adult sticklebacks of both species and non-spined fish of approximately the same size as adult Three-spined sticklebacks, or slightly larger (Fig. 137). The pike and perch varied from 10 to 25 cm; naturally, their size increased during the course of the tests.

All tests were made in the laboratory in aquarium tanks, varying, unless stated otherwise, from 60 × 35 × 35 cm to 125 × 50 × 35 cm. Hoogland's experiments were generally made in sparsely planted tanks; most of Morris's tests were done in bare tanks. In some tests the tanks were screened from their neighbours. In almost all tests the predators were kept singly. Experimental procedure was different from one test to another; details will be given with the individual tests.

As will be seen below, the details noted differed considerably from one test to another. In all tests information was obtained whether a prey fish was eaten or not, and these data were considered sufficient for the purpose of our primary ecological problem. In some tests,

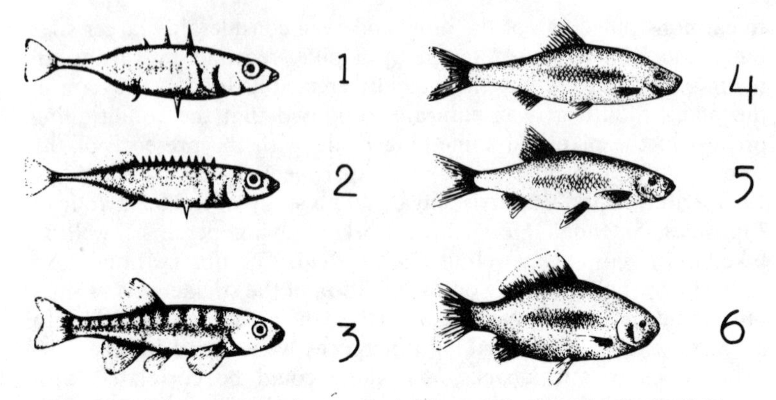

Fig. 137. The prey species used. 1 Three-spined stickleback; 2 Ten-spined stickleback; 3 minnow; 4 roach; 5 rudd; 6 Crucian carp.

particularly those of Hoogland, more detailed observations were made on the pikes' behaviour, and these throw some light on the nature of the modifications in the pikes' behaviour in the course of the tests.

Hoogland's pike showed signs of positive conditioning to the experimenter's movements prior to the introduction of the fish. Morris's predators showed escape tendencies at the beginning of some tests. Since in both cases the disturbing factor influences experiments and controls in the same way, no attempts were made to eliminate them entirely. In some of Morris's tests, where a large number of prey fish of various kinds were left with the predator for weeks on end, neither of these disturbances played a part.

3. The Reactions of Sticklebacks to Pike and Perch

Neither species of stickleback shows any tendency to escape from perch or pike as long as the predators stay motionless or move only very slowly. Quick movements of either pike or perch, particularly when directed towards the sticklebacks, evoke immediate flight, and subsequent hiding in cover. If a pike really dashes towards a stickleback in a serious attempt to capture it, the stickleback's chances of escape are small; it is never able to approach, let alone to exceed the speed of an attacking pike. A perch is much inferior in speed, and both sticklebacks have a chance to escape even from a hunting perch, as long as the latter is alone. However, in the limited space of our tanks, even single perch could easily corner and capture any stickleback. Sticklebacks may manage, by a quick avoiding movement, to make even a pike miss them, and then may escape into cover. In our tests, even in well-planted aquaria, such a frightened stickleback never stayed in cover for long, and if a pike attempted to catch it, it always got hold of it eventually. Negative results indicate almost always unwillingness of the predator, and in only very few cases (which will be indicated in the protocols) was the escape of a stickleback due to its keeping completely motionless.

The escape by a stickleback can be released by visual stimuli alone, for we have seen sticklebacks fleeing from a pike in a neighbouring tank. In spite of several attempts, following the method used by von Frisch (**4**) we could never see any evidence of a chemical response to a substance released by the damaged skin of another stickleback.

When a stickleback is seized by a pike, it reacts by erecting all its spines, keeping quite motionless except for breathing. Hoogland (**8**) has shown how *Gasterosteus* can lock its spines in this position, which enables it to keep them raised for hours on end without muscular exertion.

Gasterosteus aculeatus possesses three unpaired dorsal spines in front of the dorsal fin, one small spine in front of the anal fin, and a pair of large ventral spines. The dorsal spines are modified parts of the dorsal fin. The two anterior ones are quite large, about 6 mm in an adult fish of 65 mm; the third one rarely exceeds 1 mm in length. The unpaired anal spine, which is a modified part of the anal fin, measures also about 1 mm. The paired ventral spines, which are modified pelvic fins, are, in animals of the same size, approximately 9 mm in length. When fully raised they stand out in line with each other at right angles to the body axis; together they form one strong, rigid and very sharply pointed cross-bar of approximately 20 mm. The two foremost dorsal spines, and to a lesser extent the ventral spines, usually bear a row of small teeth along their edge.

The response of keeping motionless and raising all spines can be elicited by strong mechanical stimuli. When water is sucked from the tank through a rubber hose and a stickleback happens to be firmly caught in the hose's mouth, it will immediately show this response. When the suction is only weak, it will struggle and usually get free.

Ten-spined sticklebacks respond in the same way as Three-spined. Their spines however are all much smaller (Fig. 137).

4. The Predatory Behaviour of Perch

The hunting behaviour of the perch has been described in some detail by Deelder (**3**). He reports that in nature it is necessary for perch to co-operate with one another if prey fish are to be caught.

However, under the experimental conditions used in the present study, involving unplanted aquaria of moderate size, hunting was reduced to a minimum and the prey had no chance of escaping unless the predator rejected or avoided them. The biting of the prey by the perch is reported by Deelder (**3**) as usually being aimed at the head, and this we can confirm.

A hungry perch suddenly seeing a prey fish may, however, take it tail-first on occasions. If it is a minnow, it may be swallowed tail-first, but if it is a stickleback, it will almost always be spat out again and then, usually immediately, re-bitten head-first and perhaps swallowed. The perch has no other, more specialised, method of turning its prey round without letting go of it. In this respect it contrasts with the pike.

When it has cornered a prey fish, the perch may attempt to edge round it towards the head. The prey usually responds to these movements by turning also, so that its tail points towards the perch, and these manoeuvres may continue for some seconds. Sticklebacks under such circumstances have all their spines fully erected. Sometimes they may tilt their ventral surfaces towards the hesitating perch.

A minnow which had been bitten head-first usually vanished except for its tail. This was then frequently seen to disappear slowly and smoothly into the now closed mouth of the perch. A stickleback which had been bitten and rejected a number of times before being eaten or abandoned was sometimes held in the mouth for several minutes and sometimes released as soon as the spines touched the inside of the predator's mouth. Often, after an instantaneous rejection, the stickleback was quickly bitten once more and this was frequently seen to occur over and over again. It seemed as if the perch could not resist the visual stimulus offered by the stickleback outside its mouth, despite the fact that the mechanical stimulus of the spines always resulted in a rejection.

On a number of occasions a completely swallowed stickleback was regurgitated alive after some minutes. The perch appears to damage its prey very little, if at all, when biting it. This may well be adaptive, as certain prey species give off a chemical substance when damaged, the presence of which in the water warns nearby members of the same species of danger (von Frisch (**4**)).

After eating, the perch frequently gulps, gapes, or belches, and we have even seen the whole body vibrate with rapid jerks after a stickleback had been swallowed. The gaping movement is seen in a very exaggerated form usually just before a regurgitation. Sometimes a half-swallowed stickleback that has become lodged in the throat of the perch may be dislodged by a combination of gaping movements and vigorous sideways shaking of the head. In one instance, when a *Gasterosteus* had been half-swallowed in a single gulp by a hungry Perch, blood began to pour from the latter's gills. It swam slowly around the tank leaving a trail of blood behind it until all the tank-water was cloudy. But despite this it was the perch and not the stickleback that survived in this case.

5. The Predatory Behaviour of Pike

Apart from a few occasional observations in the wild, and from some observations on large pike in captivity, most of our knowledge of predatory behaviour of pike is based on laboratory observations on our smaller pike.

Locomotion in the pike is carried out by fin movements, by movements of the trunk and the tail, or by movements of both fins and trunk. Before describing these locomotory patterns, the observed fin movements will be described separately.

The pectorals usually perform undulatory movements, in which the fin rays move back and forth in alternation, with a slight phase difference between the neighbouring rays. The wave runs from the frontal-median ray towards the caudal-distal edge. The forward movement of each ray may be as strong as the opposite movement. In this case the fins are kept in a plane at right angles to the body axis, and the result is a water current directed downward (Fig. 138). The fins may also be directed forward, and then the forward movement of each ray is stronger than the backward motion, and the resulting water current is directed forward (Fig. 138, B). When the fin is turned back, the backward motion of the rays is stronger than the opposite movements, and the water current is directed backward (Fig. 138, C). These currents of course have roughly the opposite effect on the fish's movements.

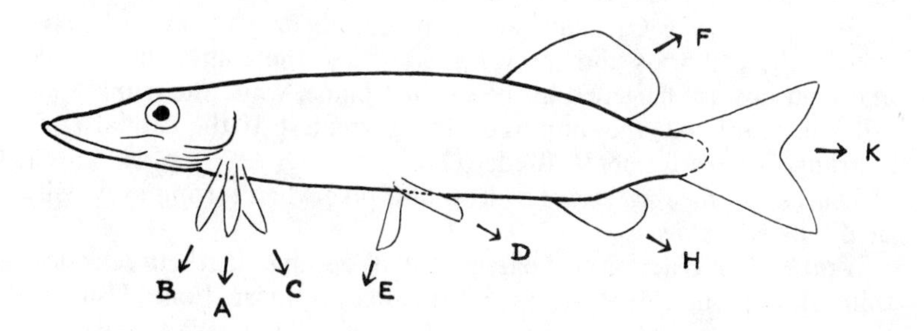

Fig. 138. Types of in movement in pike; explanation in text.

The pelvic fins are usually directed backward. They undulate also, and effect a downward and backward current (Fig. 138, D). They can be switched forward, and when undulating in this position effect a downward and forward current (Fig. 138, E).

The dorsal fin, particularly its caudal part, can undulate regularly, the waves usually travelling backward. The result is an upward, backward current (Fig. 138, F). A synchronous sideways movements of all rays of the caudal part of the dorsal fin is used in turning. The same movements can be carried out by the anal fin (Fig. 138, H); its movements are usually less pronounced than those of the dorsal fin.

Finally, the caudal fin, if moving, also undulates (Fig. 138, K); the amplitude of the fin rays' pendulum movements decreases from dorsal to ventral. The rays can also be moved synchronously.

The most common co-ordination patterns are:

Standing. The pectorals move as in A and B, the dorsalis as in F. The A type and the vertical components of B and F keep the fish at the same level and in a horizontal position; the horizontal component of B counteracts the forward impulse effected by the breathing movements. Sometimes these movements are corrected by movements of the other fins or by other movements of the same fin.

In *slow forward motion* (stalking a prey, for instance) the pectorals (C) and the dorsalis (F) are used. When speeding up, the pike uses the ventralis (D), the analis (H) and the caudalis (K) as well.

In *swimming backwards* the ventrales are used according to E, the pectorals as in B. Sometimes the movement is also controlled by the dorsal, anal and tail fins. Slowing down of forward motion is effected primarily by the ventrals (E).

In *turning* the pectorals move in opposite ways; for instance, the left pectoral may move as in B, the right one as in C. These movements are usually supported by sideways movements of dorsalis, caudalis and analis, while the trunk is curved.

When *swimming fast* (such as when 'leaping' at its prey) the pike presses the pectorals and the ventrals against the trunk, the dorsal and anal fins are flattened in the median plane, while the trunk and tail are undulated, the amplitude being greatest at the caudal end ('carangiform movement', Breder (**2**)).

The *sudden forward thrust* or leap of a pike when seizing prey will be discussed below.

A pike's first reaction to a passing fish of optimal size (between one-third and one-half of the length of the pike) is a visual one. Usually the prey is first seen with one eye. This causes the pike to turn its eyes towards the prey. This is followed by a slow turning of the whole body, which comes to a stop when the prey is exactly in front of the pike's head. In this position the prey is fixated binocularly. The pike then begins to swim forward very stealthily ('stalking'), using only the fins. At about 5 cm from the prey (in pike of the size studied) this stalking motion stops. The perception of a fish from a distance greater than the 'leaping distance' (see below) is entirely visual. According to Wunder (**10**) a blinded pike never reacts to a fish which is further away. The approach itself is also guided by visual stimuli.

Fig. 139. The S-posture of the pike assumed immediately prior to leaping at prey.

Before the pike strikes, it curves its body in an S-form (Fig. 139). This S-posture is the same as the initial posture of the carangiform movement; the frontal part of the body is less sharply curved than the caudal part. The tail is brought in a position almost at right angles to the trunk. From this position the pike, by one powerful stroke of the tail, jumps towards its prey. Under normal conditions the final curving and subsequent stretching of the tail takes place so quickly that the eye can scarcely follow the movement. If the pike's movements are hampered by water weeds, however, he may stand in the S-posture for some seconds.

In the final leap, the mouth is kept closed until just before the prey is reached. Then it is opened with force so that the prey is sucked into the mouth, as has already been described by Höller (**7**). The water is removed through the gill slits; this movement is so strong that water weeds, if accidentally taken in with the fish, slip out immediately through the slits. Once we even observed that part of the prey appeared behind the operculum.

The dash towards the prey is released by a combination of visual and mechanical stimuli. Proof of the influence of visual stimuli is given by reactions to prey or a prey substitute behind glass. The effectiveness of mechanical stimuli has been demonstrated by Wunder (**10**). Blinded pike react by jumping and snapping at moving prey at a distance of, at the most, 10 cm. It is also possible to elicit the reaction by directing a water current at the pike's head. Probably the well-developed lateral line organs on the head are responsible for this.

Once the prey is in the mouth, it is practically impossible for it to get out, mainly because of the long sharp teeth that are directed backward. A small prey is swallowed without any difficulty; a large one is first turned by short, jerky, shaking movements of the head, until it can be swallowed head-first. According to Wunder, the tongue plays an active part in this. Swallowing is effected by movements which resemble exaggerated breathing movements. These movements may continue with shorter or longer intervals until well after the prey has disappeared from sight. This may well have the function of washing the mouth (at least, one often sees loose scales being thrown out through the gill slits).

When a prey is regurgitated, which we saw several times with sticklebacks after they had been swallowed, the mouth is opened wide, and by violent jerky movements the prey is brought forward. We could not state with certainty how this was effected; possibly the tongue plays a part here as well.

Summarising, the prey-catching behaviour of a pike is a sequence of activities which could be named as follows.

1. Eye movements towards the prey
2. Turning towards the prey
3. Stalking
4. Leaping
5. Snapping
6. (Turning the prey head-forward)
7. Swallowing

Leaping and snapping are distinguished because, as will be seen, a Pike may leap at a stickleback without attempting to seize it.

6. The Responses of Perch to Sticklebacks and Minnows

Two simple tests were carried out to ascertain the reaction of two perch to a mixed group of prey species. A larger tank (210 × 36 × 36 cm) was used for this purpose. It was unplanted, and contained two predators. A known number of prey was introduced into this tank, and from then on, the number of prey surviving was scored at 24-hour intervals. One test was taken with 20 *Gasterosteus* and 20 minnows. The results are given in Fig. 140. It is seen that *Gasterosteus* was not eaten after one on the first day, and then until all minnows had gone. In another test, 20 *Gasterosteus*, 20 *Pygosteus* and 20 minnows were supplied at the start. Fig. 141 shows the result. *Gasterosteus* was left alone until more than half of the other two species had disappeared. Until the 7th day the minnows were preyed on slightly more than *Pygosteus*; after that, *Pygosteus* was taken mainly. The difference between *Gasterosteus* and minnows is striking in both cases.

The vast majority of observations concerned the responses of a single perch to a single prey fish. The various species of prey fish were given in different sequences. Before giving the results obtained from using special sequences, the total results from all these tests,

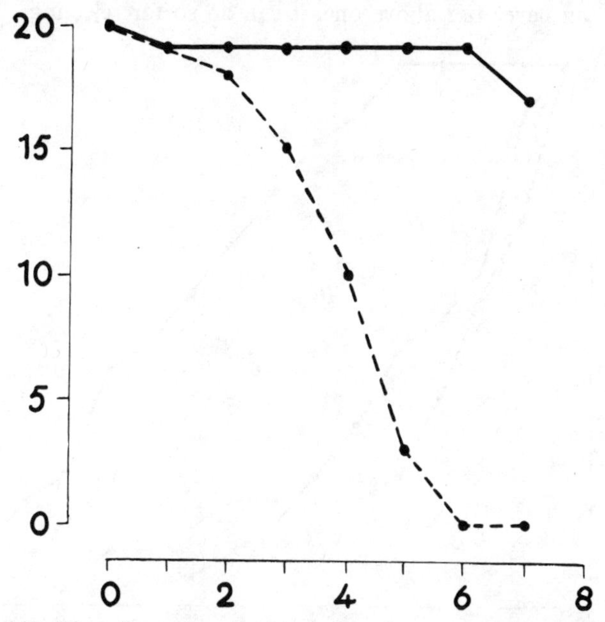

Fig. 140. The effect of predation by perch on a mixed shoal of 20 *Gasterosteus* (solid line) and 20 minnows (broken line). Ordinate: numbers of prey present. Abscissa: time in days.

taking each presentation singly, is recorded in Fig. 142. In all, 475 such presentations were made to 5 perch. At the outset it was hoped that the tests could be standardised in respect of the length of time during which the prey was available to the perch, but this desirable method was found to be unsatisfactory, because the time lag until the occurrence of the first response varied erratically, and this could not be correlated with the nature of the response that followed it. However, since the prey was always either eaten very soon after its introduction or, if not eaten, was completely ignored after a period of 10–20 minutes, tests were concluded either when the prey had been eaten or when the perch had lost interest in it.

In Fig. 142 the average number of reactions per test is given for each of the three species when presented to perch. The predator's reactions have here been classed as eating, rejecting, and avoiding. It ought to be stressed that, although a prey fish can only be eaten once per presentation, it can be rejected a number of times. A perch may bite and then spit out a fish repeatedly and each time it is spat out it is recorded as one rejection. Thus, although the ordinate value for eating can never rise above one, it can do so for rejecting. This is

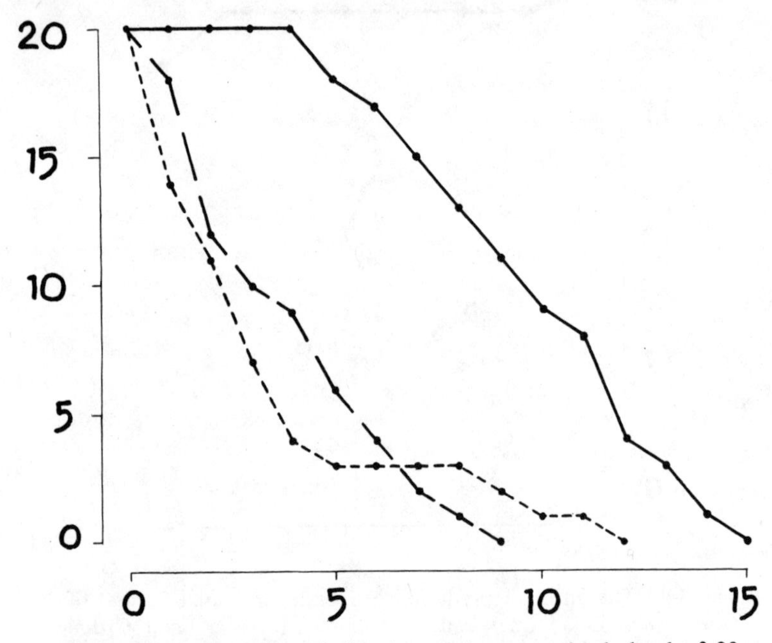

Fig. 141. The effect of predation by perch on a mixed shoal of 20 *Gasterosteus* (solid line), 20 *Pygosteus* (broken line), and 20 minnows (dotted line). Ordinate: numbers of prey present. Abscissa: time in days.

also true of avoiding. An avoidance was scored each time a perch made an intention movement of feeding, but stopped before actually making contact.

Fig. 142 shows clearly that the minnow is more frequently eaten and less frequently rejected, or avoided, than *Pygosteus*. Further, *Pygosteus* is more frequently eaten and less often rejected or avoided than *Gasterosteus*. Also, the difference between *Gasterosteus* and *Pygosteus* is much more striking than that between *Pygosteus* and the minnow.

Part of the above results were obtained by giving the prey fish in the following sequence: 1. *Gasterosteus*, 2. *Pygosteus*, and 3. minnow. As soon as the predator had lost interest in, or had eaten, the *Gasterosteus*, this prey was, if necessary, removed, and a *Pygosteus* was introduced. Similarly, when the predator had eaten (or lost interest in) this, it was removed if necessary, and a minnow introduced.

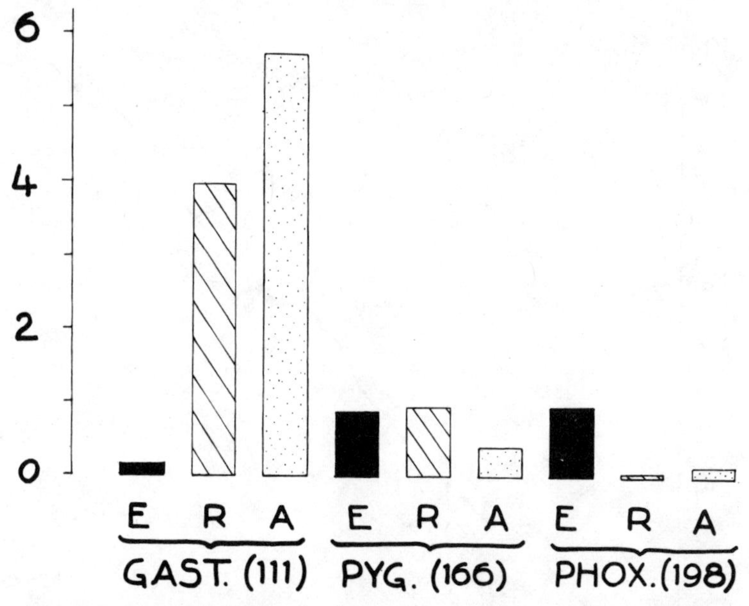

Fig. 142. Results of 475 single presentations of 111 *Gasterosteus*, 166 *Pygosteus*, and 198 minnows to 5 perch. Ordinate: average number of responses per presentation. Abscissa: responses classified for prey species and for eating (E), rejecting (R) and avoiding (A) by perch.

In all, 64 such test sequences were carried out. There were four possible ways in which the sticklebacks could be treated in each series, if the predator's response was simply recorded as Eating (with or without preceding rejections) or Not Eating. The frequency in which these four sequences occurred were:

Gasterosteus not eaten, *Pygosteus* eaten	56	*Gasterosteus* eaten *Pygosteus* not eaten	1
Gastereosteus not eaten *Pygosteus* not eaten	7	*Gasterosteus* eaten *Pygosteus* eaten	0

It should be noted that in all these 64 cases the minnow, presented afterwards, was eaten.

Another set of sequential tests was made with three 'novice' perch. These three had never had any experience of sticklebacks in their lives, since they were caught in an isolated pond where sticklebacks were absent. With them it was possible to observe the manner in which they modified their feeding behaviour when presented daily

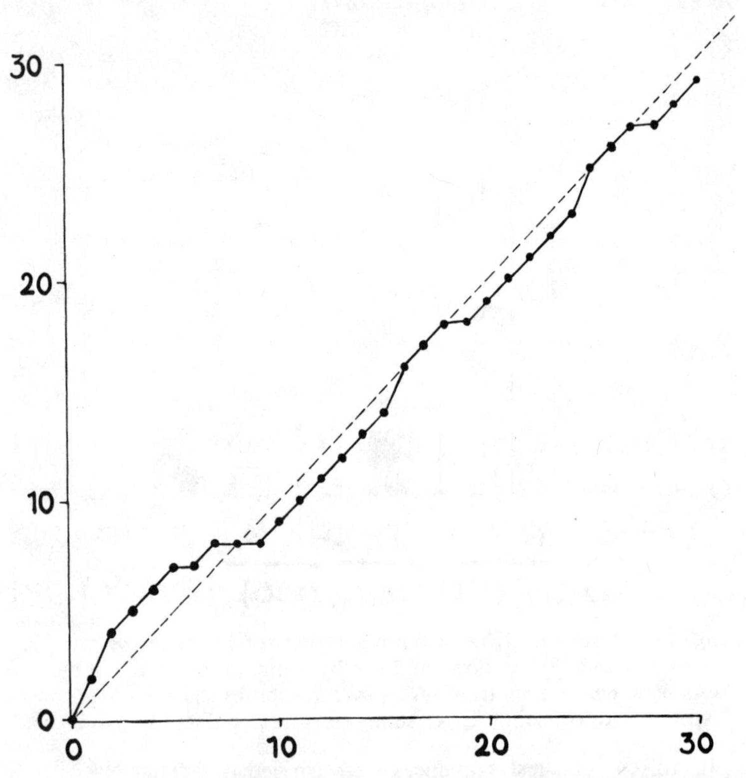

Fig. 143. Records of responses to *Gasterosteus* shown by 'novice' perch no. 1 over a period of 30 days (abscissa) showing no modification with experience. Ordinate indicates total score to date. Scoring method: 0 = avoid; 1 = bite and reject; 2 = eat.

with one stickleback followed by one minnow. An attempt was made to keep the amount of food they ate each day as near constant as possible, by only giving a minnow if a stickleback was not eaten.

As *Gasterosteus* was available in larger numbers than *Pygosteus* at the time, two of the perch were given the former and one the latter. Figs. 143, 144 and 145 show the way in which these three perch modified their behaviour with repeated testing at 24-hour intervals over a period of about a month each. The graphs are summative; when they rise above the dotted line they indicate increased response to the sticklebacks, when falling below they show growing avoidance. Fig. 143 shows that this perch did not modify its behaviour with respect to *Gasterosteus* during the experimental period. The second

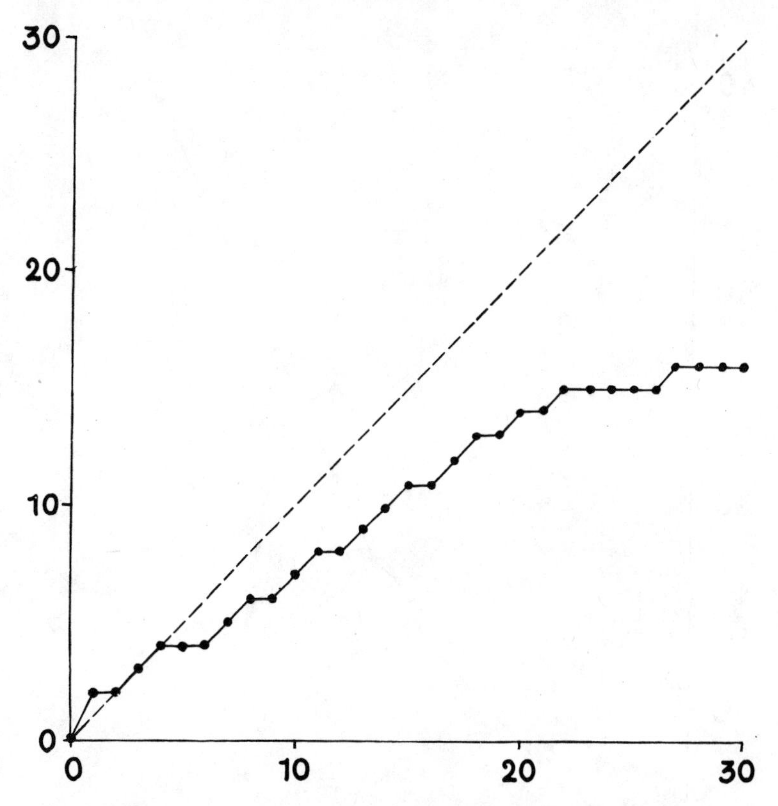

Fig. 144. Records of response to *Gasterosteus* shown by 'novice' perch no. 2 over a period of 30 days (abscissa) showing the development of avoidance. Explanation as for Fig. 143.

perch (Fig. 144) however, did; although it always ate its minnow, it learned to avoid the stickleback more and more. The third perch (Fig. 145) actually learnt to eat *Pygosteus*. It is possible that these perch were kept at a semi-starvation diet; pike of the same size seemed to thrive best on a ration of three to four times as much as that offered to these perch (see later). This last perch was given, after it had learnt to eat *Pygosteus*, a month of *Gasterosteus* presentations, and like the perch of Fig. 144 it too learnt to avoid this species.

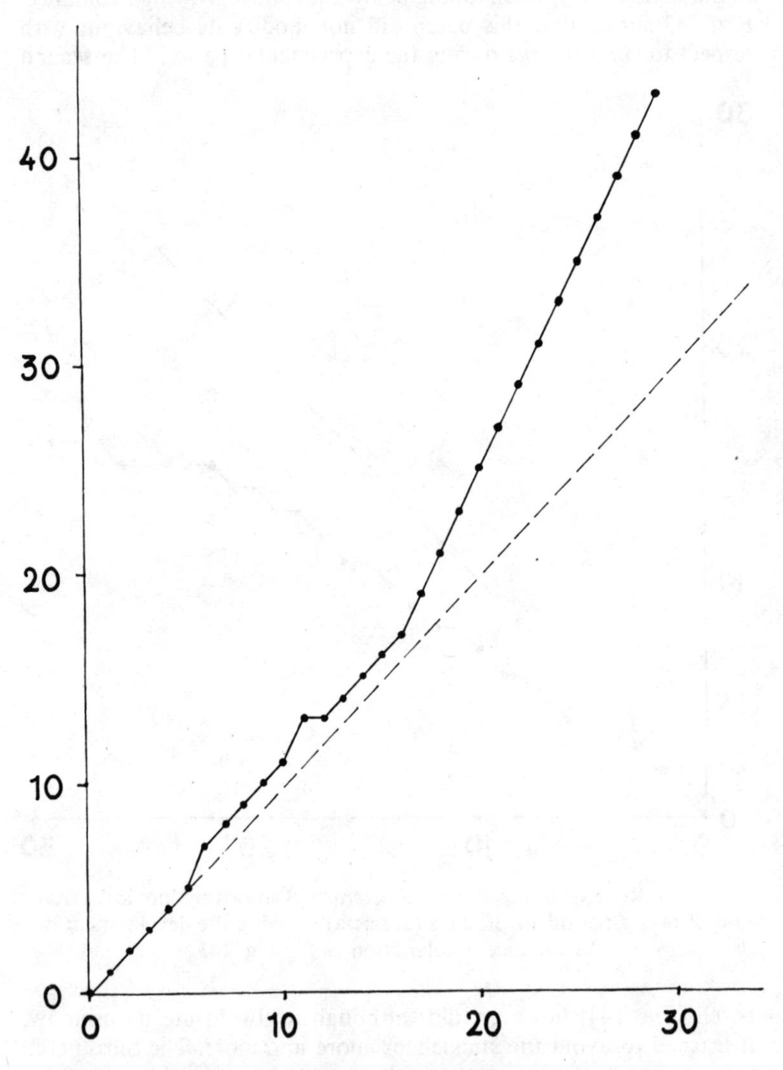

Fig. 145. Records of responses to *Pygosteus* shown by novice perch no. 3 over a period of 30 days (abscissa) showing development of eating. Explanation as for Fig. 143.

7. The Responses of Pike to Sticklebacks and Other Fish

The observations were made mainly on eight pike. Five were used in Hoogland's tests; they were kept separately in tanks with open vegetation. Morris used three pike which were kept in tanks bare of vegetation. Hoogland usually watched the pike's behaviour until the prey offered was eaten, or, if it was not eaten, until 30 minutes after it had been presented. Morris broke off his observations after the prey had been either eaten, or when the pike lost interest.

All pike gave chase to a stickleback when it was first presented. In these first tests the stickleback was always caught as promptly as any other fish of the same size. Usually it was rejected immediately after it had been snapped up, and the pike would make violent coiling movements, and would 'cough' intensely several times. The stickleback would swim away into cover, but it soon calmed down, and began to move about again. This might cause the pike to catch it again, and it might then try to swallow it. In this it rarely succeeded, and if so, it took a long time. The stickleback, keeping entirely motionless with spines raised, was manoeuvered into various positions by violent head movements of the pike; sometimes it would almost disappear from sight only to be shifted back forward again, in any position. The drawings of Fig. 146 were made from Tinbergen's film.

This might go on for over an hour without the stickleback being either swallowed or even killed. In most cases it was finally rejected, sometimes dead, but more often alive. In other cases it was eventually eaten.

As with the perch, some tests were done with mixed groups of prey species. In one experiment, 2 pike were offered 12 *Gasterosteus* and 12 minnows. As shown in Fig. 147 the minnows were eaten first, and no *Gasterosteus* was eaten until all minnows had gone. Again, the sequence was not recorded in detail, so that it is uncertain whether the first stickleback was eaten after all minnows had gone, or a little earlier on that same day.

Fig. 148 shows the results of a test in which 12 *Gasterosteus*, 12 *Pygosteus* and 12 minnows were offered simultaneously. Apart from one *Gasterosteus* and one *Pygosteus* eaten on the first day, both species were left alone until, on the 5th day, all minnows had been eaten. The pike then attacked both species of sticklebacks, but *Pygosteus* was slightly preferred.

In other tests, various fish were presented singly to a single pike, and the pike's behaviour observed. In Figs. 149 and 150 the results are given as averages of the three types of response scored: Eating,

Rejecting, and Avoiding. Again, as in the tests with perch, Rejecting and Avoiding might be observed more than once during a test, whereas Eating could only be scored once.

The figures for Rejecting were on the whole higher in perch than in pike, because the perch spits out a fish in order to turn it round, and

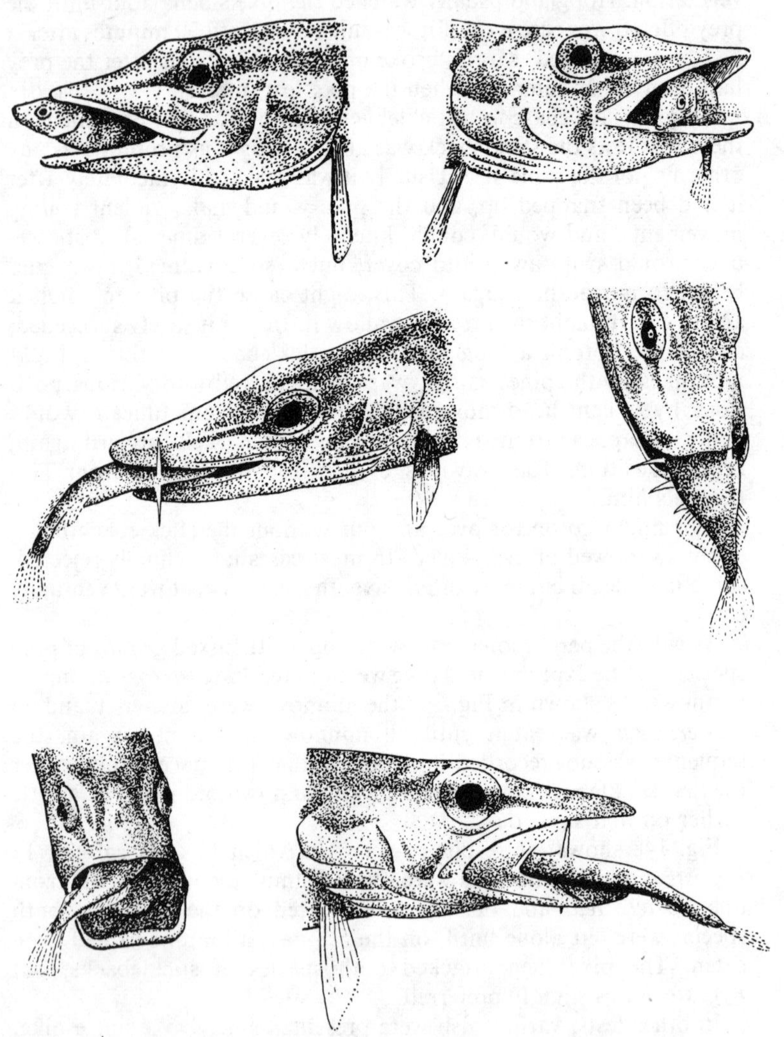

Fig. 146. Drawings taken from 16-mm. film showing the variety of positions of *Gasterosteus* in the mouth of the pike.

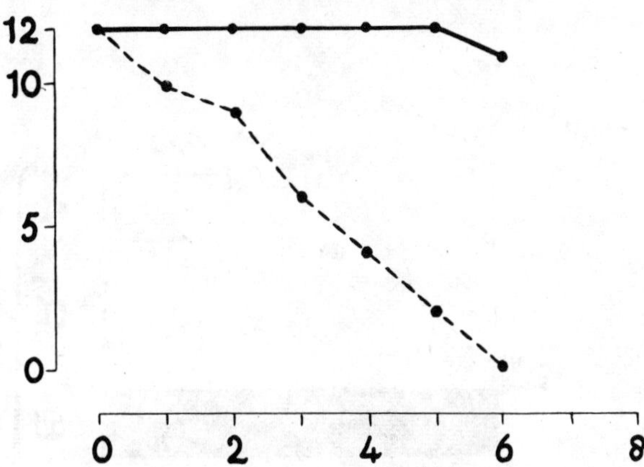

Fig. 147. The effect of predation by pike on a mixed shoal of 12 *Gasterosteus* (solid line) and 12 minnows (broken line). Ordinate: number of prey present. Abscissa: time in days.

this was recorded as a reject. The pike turns a prey fish without allowing it to leave its mouth. Avoidance is also much more easily seen in perch than in pike, because a perch will often swim some distance

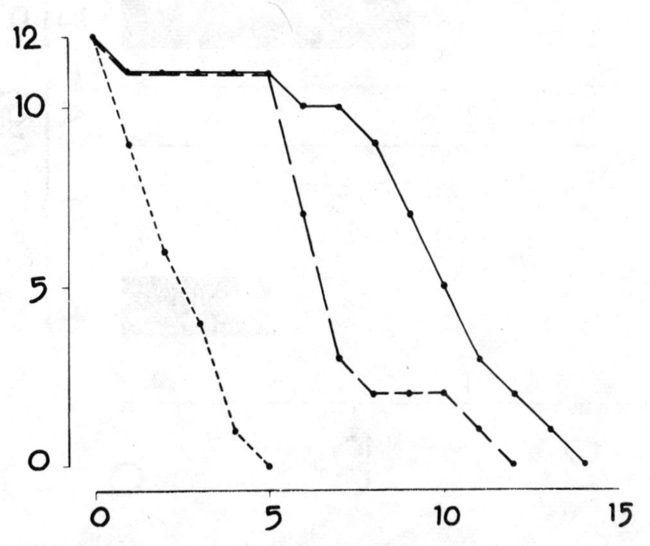

Fig. 148. The effect of predation by pike on a mixed shoal of 12 *Gasterosteus* (solid line), 12 *Pygosteus* (broken line) and 12 minnows (dotted line). Ordinate: number of prey present. Abscissa: time in days.

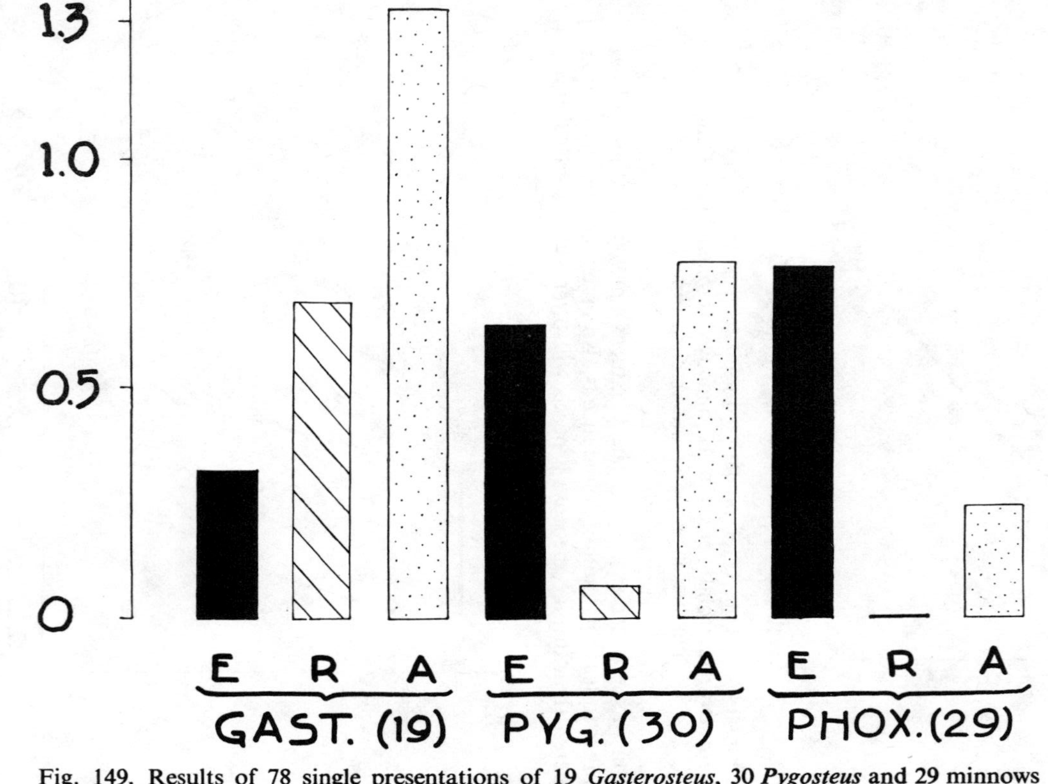

Fig. 149. Results of 78 single presentations of 19 *Gasterosteus*, 30 *Pygosteus* and 29 minnows to 3 pike, scored as in Fig. 142.

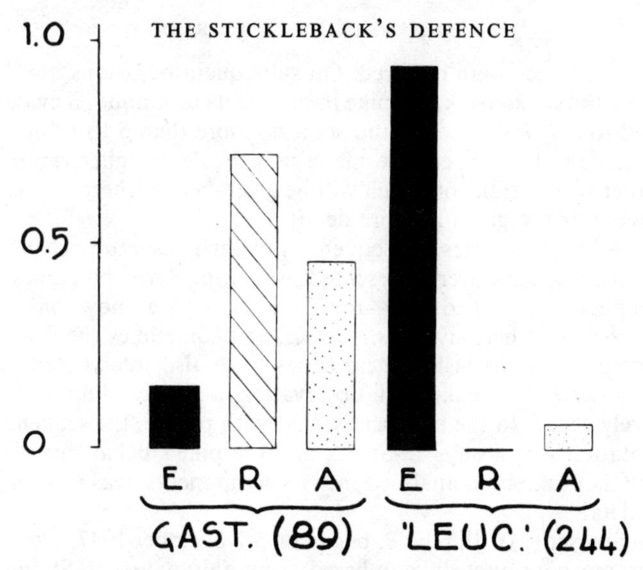

Fig. 150. Results of 333 single presentations of 89 *Gasterosteus* and 244 '*Leuciscus*' to 3 pike, scored as Fig. 142.

towards a prey before giving up, while a pike may merely turn its eyes towards a prey without even turning its body, a response that we may have missed in some of the series. The figures for perch and pike are, therefore, not entirely comparable, but both are consistent in themselves, and are reliable as evidence on our main problem.

Fig. 149 summarises Morris's results obtained with 3 single pike, and *Gasterosteus*, *Pygosteus*, and minnow as prey species (78 presentations in all). Fig. 150 gives Hoogland's results with three single pike (E, B, and D) which were usually offered one or more specimens of the three non-spined species mentioned above, followed by one *Gasterosteus* and then again followed by one fish of the non-spined group. There were 333 presentations in all.

Pike therefore respond in roughly the same way as perch: they reject and avoid *Pygosteus* more than minnows or other non-spined fish, and *Gasterosteus* more than *Pygosteus* and much more than non-spined species.

Not summarised in Figs. 149 and 150 are the results obtained during a pilot test in which a pike was given one *Gasterosteus* two or three times a week, and no other fish was given. At the temperature prevailing during these experiments this was a starvation diet. Our notes on this series are incomplete, but the following information is available: at the first presentation it took the pike about 30 minutes to swallow the stickleback entirely. On the second occasion again 30 minutes were required. On subsequent occasions, far from rejecting the stickleback, the pike improved its technique; it evidently learned to eat *Gasterosteus*, and soon no more than 5 to 10 minutes were needed. This experience made us provide a higher ration for the other pike, details of which will be given below where the results of three tests are given in more detail.

In the first two series the sequence in which *Gasterosteus* and the non-spined species were presented was irregular, sometimes the non-spined fish (called 'L.'—for *Leuciscus*—from now on) were given first, on other days *Gasterosteus*, and sometimes the fish were presented simultaneously. These series were also interrupted occasionally, and the method of observation and recording was still relatively crude. In the final series, that with pike E, the sequence of presentation was always L.-St.-L., and the pike's behaviour, which had by that time been analysed into its components, was recorded in more detail.

The first series, with pike B, began on 5 November 1947. The daily entries are continuously numbered from this date on. St. means stickleback (*Gasterosteus*), P. means pike, L. means non-spined prey as specified above.

1. 1 St. caught at once, swallowed until invisible but regurgitated at once. After this P. shows occasional interest in St. (which is not very active and dies after 20 min); twice P. jumps without snapping.
2. 1 St. captured immediately but released at once; after this P. shows continuous interest. After 3 min, St. seized again; St. remains in mouth of P. for 2 min, then released. During remaining 25 min St. moves very slowly, P. shows occasional interest. Then L. given, P. responds weakly, but snaps and eats it after 13 min.
3. 1 St., not very active. P. shows little interest. 1 L. which was given together with the St. is attacked 3 times. After two misses, P. eats it.
4. No test, P. not fed.
5. No test, P. not fed.
6. 1 St. caught at once, is eaten after 18 min struggle. After this 1 L. is eaten at once.
7. 1 St.; P. mildly interested. Not captured. 1 L. given afterwards, eaten almost at once.
8. 1 St.; P. shows little interest.
9. Pike has regurgitated 2 L. No test (on days 7, 8 and 9 ventilation of tank was insufficient).

10. 1 L. eaten at once. 1 St. captured at once, kept in mouth for 65 min, finally released (alive). After this 1 L. is eaten at once.
11. 1 St. caught at once, released after 10 min. P shows occasional interest and seizes St. again after 10 mins, releases it, captures it again, swallows it after 25 min. 1 L. is then given and eaten at once.
12. No test.
13. 2 L. eaten at once. 1 St. captured at once, swallowed after 10 min.
14. 2 L. eaten at once. 1 St. seized immediately, released after 2 min; after this occasional interest, no more.
15. 2 L. eaten a tonce. 1 St. seized immediately, released after 4 min, seized again, swallowed after 15 min.
16. 2 L. eaten at once. 1 St., captured and released three times in quick succession, after that P. shows only occasional interest.
17. 2 L. eaten at once. 1 St. captured at once, released after 7 min, seized again, swallowed after 3 min.
18. 2 L. eaten at once. 1 St. (relatively inactive) captured after 15 min but released at once. P. remains mildly interested. Then 1 L. eaten at once.
19. No test.
20. 1 L. eaten at once. Then 1 St. and 1 L. offered simultaneously. Although St. is the more active of the 2, P. shows more interest in L. and eats it after 30 min.
21. No test.
22. 2 L. eaten at once, 1 St. seized at once, released after 14 min.
23. 2 L. eaten at once; no interest for St.; another L. given, not eaten after 30 min.
24. 3 L. eaten at once. 1 St. releases jumping without snapping; St. is seized after 20 min and released at once.
25. 2 L. eaten at once.
26. No tests.
27. 2 L. eaten at once. 1 St. seized and released twice. Seized again after 5 min; after another 5 min it is eaten.
28. 2 L. eaten. 1 St. caught at once, swallowed after 15 min.
29. 2 L. eaten. 1 St. caught and released twice, seized again and swallowed, then rejected, finally eaten.
30. 2 L. eaten at once. 1 St. seized at once, swallowed after 5 min.
31. 3 L. eaten. 1 St. (not very active) almost ignored; another L. is eaten at once.
32. 3 L. eaten at once. 1 St. is seized after 20 min and released. 1 L eaten at once.
33. No test.

34. 2 L. eaten at once; 1 St. eaten almost at once.
35. 3 L. eaten at once; 1 St. captured at once, swallowed after 15 min. Another L. is not eaten.
36. Yesterday's St. has been regurgitated; spines are still raised. 2 L. eaten at once. 1 St. captured and released, after than P. shows occasional interest. A new L. is eaten at once.
37. 1 L. eaten.
38. 2 L. eaten.
39. 1 large L. eaten. 1 St. releases jumping without snapping. P. shows only mild interest after this. Another L. is eaten at once.
40. No test.
41. 1 L. eaten at once. 1 St. evokes only mild interest. Another L. is eaten at once.
42. 2 L. eaten at once. 1 St. is almost ignored. A new L. is eaten at once.
43. 2 L. eaten at once. 1 St. seized at once, released after 3 sec, then ignored. A new L. is eaten at once.
44. 2 L. eaten at once. 1 St. releases jump without snapping; later it is snapped up and released at once. P. loses interest. Another L. is eaten at once.
45. 2 L. eaten. Tests discontinued.

In this series, the following points are brought out: Of 75 L. offered, 73 were eaten. Of 32 St. offered, 11 were eaten. Of 21 St. that were not eaten, one was killed, and 20 survived. Of these, 9 were never even captured at all, as compared with 2 out of 75 L.

In comparison with pilot experiments, not reported here in detail, in which two pike learnt, after a few trials, not to capture any *Gasterosteus* at all, the series of pike B shows few signs of this kind of learning: only 9 out of 32 were completely avoided.

The high proportion of sticklebacks eaten or at least snapped up might be due to the fact that the total diet of this pike was still relatively low. It seemed possible that the pike's behaviour was controlled by two conflicting tendencies, *viz* that demonstrated by a starving pike, which learns to eat sticklebacks, and the tendency to stop reacting to sticklebacks altogether. Therefore, another test series was done with a new pike (D) which was fed more liberally. Before starting the tests, the pike was given 4 L. a day for some weeks. It was then allowed one day to eat as much as it would accept. It took 10 L., but then refused food for the rest of the day and for the whole of the next day. After that it took 4 L. again. A ration of 3 or 4 daily was therefore considered sufficient to prevent starvation under the temperature conditions of the test.

The series of pike D then proceeded as follows:

1. (27 Jan. 1948) 3 L. eaten at once. 1 St. seized 5 times and released each time within 1–20 sec. After that P. shows interest occasionally; one jump without snapping. 1 L. eaten at once.
2. 3 L. eaten. 1 St. seized and released at once. P. loses interest. A second St. is put in and is seized at once and rejected. 1 L. eaten at once.
3. 3 L. eaten. 1 St. evokes mild response at first, is seized after 5 min but not held firmly, and released at once. After this, weak and occasional responses except once when P. jumps but fails to snap. 1 L. eaten at once.
4. No respose to L. No St. offered.
5. 3 L. eaten. 1 St. releases no response, but neither does a L. offered afterwards.
6. 3 L. eaten. Mild interest in St. 1 L. eaten at once.
7. 3 L. eaten. P. pays no attention to St., 1 L. is eaten after some delay, 2 further L. are left alone.
8. 3 L. eaten. No attention for St. except 1 jump without snapping. 2 L. eaten at once.
9. 3 L. eaten. Mild interest in St. 2 L. not eaten.
10. 2 L. eaten. 1 St. captured at once and released immediately. This happens a second time. Later one jump without snapping. 1 L. eaten at once.
11. 1 St. seized at once but released, no further attention. No L. offered.
12. No test.
13. No test.
14. No test.
15. No test.
16. 2 L. eaten. 1 St. seized at once and released without delay. One more jump without snapping. P. loses interest. L. seized and killed but not eaten.
17. 2 L. eaten. 1 St. releases one aiming response, no jump. 1 L. eaten at once.
18. 2 L. eaten. 1 St. evokes mild interest. 2 L. eaten.
19. No test.
20. No test.
21. 2 L. eaten. 1 St. seized at once, and released. P. loses interest. 1 L. eaten at once.
22. 2 L. eaten. 1 St. seized at once and released; this is repeated twice in 30 min. 1 L. eaten at once.

23. 2 L. eaten. 1 St. seized at once and released. Repeated after 8 min, then after 3 min. 1 L. is eaten at once.
24. 2 L. eaten. 1 St. evokes mild interest. 1 L. not eaten.
25. As on 24.
26. 2 L. eaten. 1 St. causes mild interest. 1 L. releases repeated jumping but no snapping.
27. No test.
28. 1 L. eaten. Only passing attention for St. No further L. offered.
29. P. regurgitates one L.; no test.
30. 2 L. eaten. 1 St. seized at once and released. P. shows mild interest for another 5 min. 1 L. eaten at once.
31. 1 large L. eaten. Then a new L. and 1 St. offered together. L. evokes much more interest than St., but neither is caught.
32. 2 L. eaten. No interest in St. A new L. is eaten at once.
33. 1 L. eaten. 1 St. caught at once, but released, and P. loses interest, snaps once. L. is eaten at once.
34. No test.
35. 2 L. eaten. Mildly interested in St. but no jump. 1 L. is eaten at once.
36. 2 L. eaten. 1 St. (not very active) evokes mild interest. A (very sluggish) L. is not caught either, though P. shows occasional interest.
37. 2 L. eaten. 1 St. causes mild interest. Another L. is eaten at once.
38. 1 L. eaten. 1 St. causes mild interest. 1 L. is caught but released.
39. 1 large L. eaten. No interest in St. Another L. is jumped at, missed, then eaten 5 min later.
40. 1 L. eaten. 1 St. is looked at once. Another L. not eaten, but it failed to move.
41. 1 L. eaten. 1 St. snapped up but released. P. loses interest. Then jumps once more, but misses. 1 L. eaten at once.
42. No test.
43. 1 L. eaten. 1 St. is seized but released, then left alone. 1 L. eaten at once.
44. 1 L. eaten. No interest in St. Another L. eaten at once.

Tests discontinued.

In this series, of 100 L. offered, 88 were eaten. Of 34 sticklebacks offered, none was eaten. Of 14 snapped at but not eaten all survived. 20 St. were never captured at all, as compared with 12 out of 100 L. In 23 cases when a stickleback was not eaten, a L. offered immediately afterwards was eaten at once. In 9 such cases the L. was not eaten either. The opposite (St. eaten, L. refused) never happened.

Series D, though carried out with a pike in a not too hungry condition, still showed some defects. First, it was evident that the behaviour of the pike was dependent to a certain extent on the degree of activity of the prey. A failure to react to the stickleback seemed occasionally to be due to the latter's entire lack of movement and not to lack of responsiveness of the pike. We decided therefore to find some measure of the stickleback's activity. In the next series, the tank was divided by imaginary planes into 16 compartments, and each time the stickleback moved from one into the other was scored as one; the added total during the test was taken as a rough measure of the stickleback's activity. Next, the notes on the nature of the pike's response could be made much more detailed and accurate. We therefore described the pike's activity in terms of partial reactions according to the schedule given on p. 60. It had been found that the behaviour could end with any activity from 1 (turning the eyes towards the prey) to 7 (swallowing) inclusive. Incomplete reactions always began with 1, and the sequence of activities was always the same. Only when the external stimuli, needed for a later link, were provided right at the beginning, could the preceding links be omitted. Thus a fish appearing exactly in the median plane within striking distance could release snapping without preceding eye movements, orientation of the body, and stalking. But in fact this rarely happened in our tests. Our scoring system was as follows: a reaction stopping after the initial eye movements was scored as 1; one ending after orienting the body was given 2; one ending after stalking received 3; one stopping after leaping got 4; one ending after snapping was scored 5; one ending after rearranging the prey in the mouth got 6; and a complete response which included eating received 7. For incomplete parts of the chain ½ was scored; thus incomplete approach was scored as 2½. In the entries given below the separate reactions in each test are given as well as their sum total, giving an idea of the total responsiveness of the Pike.

Test series with Pike E started on 19 March 1948.

1. 3 L. eaten. Activity of stickleback 114. Activity of pike: 5—4½—2—2—2—5—4—5—5—5—5—4—5—5—5—5—5—5—5—5—2—2—4—2—4—3—2—3—4—5—2—2—1½—3—2—3—2½—2½—3—5—2—2½—2—2—2½—2½ = 159½
   1 L. eaten at once.
2. 2 L. eaten. Act St. 38. Act. Pike: 4—5—2½—2½—1—1½—1½— = 18. 1 L. eaten after 5 min.
3. 2 L. eaten. Act. St. 71. Act. Pike: 2—2—2—1½—2—1—1½—3—2 = 17. 1 L. eaten at once.

4. 2 L. eaten. Act. St. 24. Act. Pike: 3—1½—2 = 6½. 1 L. refused.
5. 1 L. eaten, 1 L. refused. Act. St. 38. Act. Pike: 0. 1 L. refused.
6. 1 L. eaten. Act. St. 34. Act. Pike: 5—3—2—2—2—2—2½ = 18½. 1 L. eaten at once.
7. 1 large L. eaten. Act. St. 38. Act. Pike: 5—2½—2½ = 10. 1 L. eaten at once.
8. 2 L. eaten. Act St. 5. Act. Pike: 4—2½— = 6½. 1 L. eaten at once.
9. 2 L. eaten. Act. St. 75. Act Pike: 2—2—1½—1—1½ = 8. 1 L. eaten at once.
10. 2 L. eaten. Act. St. 34. Act. Pike: 3—1—1—5—2½ = 12½. 1 L. eaten at once.
11. 2 L. eaten. Act. St. 28. Act. Pike: 2—4—3—1— = 10. 1 L. eaten at once.
12. 2 L. eaten. Act. St. 45. Act. Pike: 4—3—3—2—2—1 = 15. 1 L. eaten at once.
13. 2 L. eaten. Act. St. 27. Act. Pike: 5—2—4—1½ = 12½. 1 L. eaten at once.
14. 2 L. eaten. Act. St. 67. Act. Pike: 2—1½ = 3½. 1 L. eaten at once.
15. 2 L. eaten. Act. St. 4. Act. Pike: 5. 1 L. eaten at once.
16. 3 L. eaten. Act. St. 31. Act. Pike: 4—3—1 = 8. 1 L. eaten at once.
17. 2 L. eaten. Act. St. 16. Act. Pike: 5—1 = 6. 1 L. eaten at once.
18. 3 L. eaten. Act. St. 20. Act. Pike: 2½—1½ = 4. 1 L. eaten at once.
19. 3 L. eaten. Act. St. 123. Act. Pike: 5—1—1—1½—2½—1 = 12. 1 L. eaten at once.
20. 2 L. eaten. Act. St. 144. Act. Pike 1½—1½—1½—1—1—2½—1 = 10. 1 L. eaten at once.
21. 2 L. eaten. Act. St. 155. Act. Pike: 2—1—1—2½—1 = 7½. 1 L. eaten after 3 min.
22. 2 L. eaten. Act. St. 108. Act Pike: 3—1½—1½ = 6. 1 L. eaten at once.
23. 2 L. eaten. Act. St. 168. Act. Pike: 1—1½—1½ = 4. 1 L. eaten after 1 min.
24. 2 L. eaten. Act. St. 5. Act. Pike: 2½—2½ = 5. 1 L. eaten after 20 min.
25. 1 large L. eaten. Act. St. 67. Act. Pike: 2—2½—2—2 = 8½. 1 L. eaten after 5 min.
26. 2 L. eaten. Act. St. 89. Act. Pike: 1½—2—2—1—4 = 10½. 1 L. eaten at once.
27. 1 L. eaten. Act. St. 9 (it dies after 10 mins). Act Pike: 3—2—2—2—2—1 = 12. 1 L. eaten at once.

28. 2 L. eaten. Act. St. 28. Act. Pike: 3½—3½—3—3—2½—1½—2—2½ = 21½. 1 L. eaten at once.
29. 4 small L. eaten. Act. St. 41. Act. Pike: 5—2—1 = 8. 1 L. eaten at once.
30. 2 L. eaten. Act. St. 19. Act. Pike: 5—2½—1 = 8½. 1 L. eaten at once.
31. 3 L. eaten. Act. St. 134. Act. Pike: 5—2½—1—2—2—1—2—1—1—1½—2½—1 = 22½. 1 L. eaten at once.
32. 4 L. eaten. Act. St. 45. Act. Pike: 5—5—5—2½—4 = 21½. 1 L. eaten at once.

In this series, of 101 L. offered, 98 were eaten. Of 31 sticklebacks offered (we exclude No 27 for obvious reasons) none was eaten. Of 13 sticklebacks snapped at, all survived. 18 sticklebacks were never captured at all, as compared with 3 out of 101 L. In 29 cases when a stickleback was not eaten, a L. offered immediately afterwards was eaten at once. In the 2 other cases the L. was not eaten either. The opposite never occurred.

Fig. 151 gives a graphic summary of series E. It will be seen that the activity of the stickleback had practically no influence on that of this pike. As long as the stickleback did not keep quite still, the pike showed some response. This response, shown again and again on level 5 and in high intensity on the first day, dropped sharply after this and remained more or less constant and at a very low level throughout the 32 days of the test. It never got beyond 5; that is to say, the pike never swallowed the stickleback. The average response; while showing some tendency to decrease even after the second day, shows some irregular peaks. However, this average was taken from the positive responses shown; the negative responses (those occasions when the stickleback was moving within view without even evoking a response of level 1) were not noted, and in fact it would be difficult to design a scoring system for them.

The occasional snapping responses to the stickleback, as recorded on days 2, 6, 7, 10, 13, 15, 17, 19, 29, 30, 31, and 32 were for the greater part due to a disturbing factor which we did not try to eliminate: the Pike became positively conditioned to the experimenter's movements when introducing a fish into the tank, and sometimes shot forward at the prey immediately on its introduction. The 'fives' scored on these 12 days were all of this type (open triangles) except 2 (those on day 10 and day 32), which are indicated by solid triangles.

8. The Responses of Perch and Pike to De-spined Sticklebacks

In a number of tests, sticklebacks were offered to perch and pike after

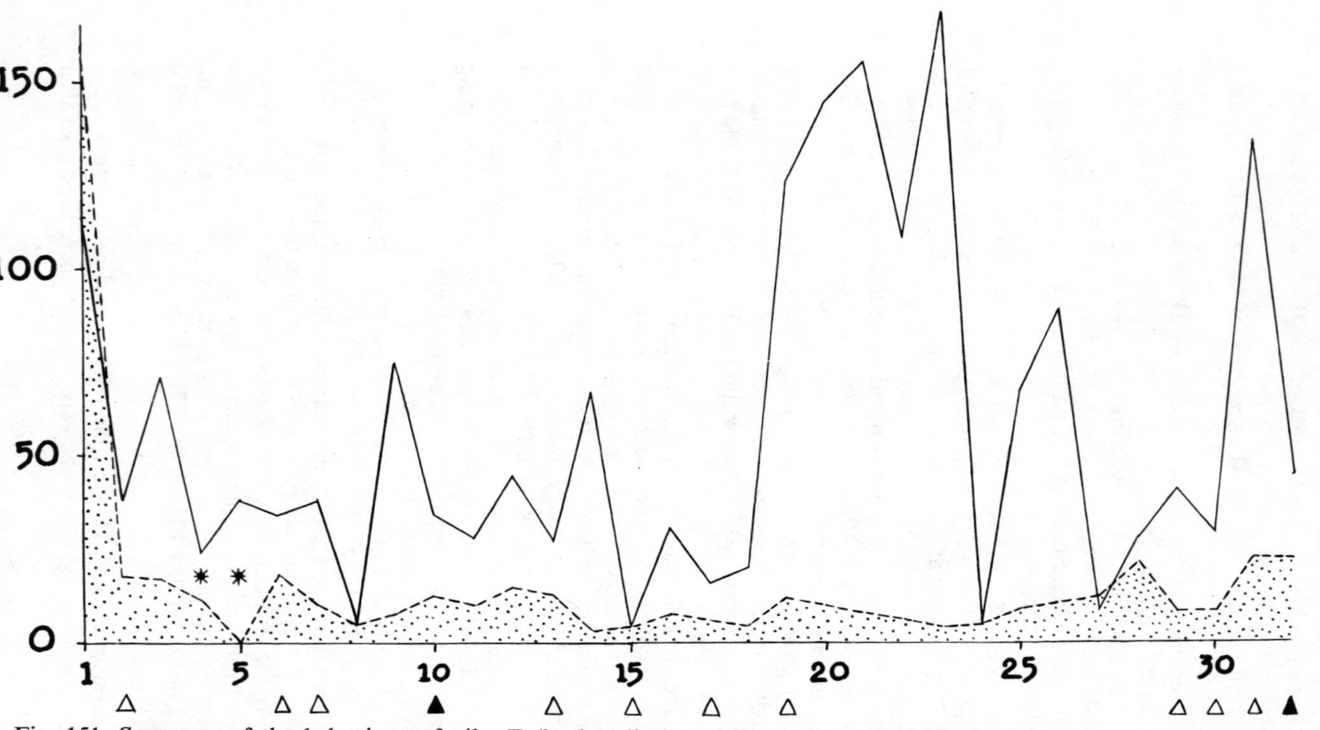

Fig. 151. Summary of the behaviour of pike F (broken line) and *Gasterosteus* (solid line) given as prey over a period of 32 days (abscissa). Ordinate: total activity score. Asterisks indicate days when non-spined prey ('*Leuciscus*') was refused. Triangles: stickleback snapped at. Further explanation in text.

their spines had been cut off. In order not to injure the sticklebacks, the bases of the spines were left intact, and the result was that 'de-spined' fish were not as smooth as the non-spined fish used; yet the main obstacles preventing the predators from swallowing sticklebacks were gone, and the remaining stumps were less sharp than the intact spines. In short series with perch and pike, involving about 150 *Gasterosteus*, such de-spined fish were almost invariably eaten. The following figures apply to two series done with pike A and C and de-spined *Gasterosteus*. Of 131 L. offered, 121 were eaten and 5 refused. Of 63 de-spined *Gasterosteus* offered, 55 were eaten and 8 refused. The slightly larger proportion of *Gasterosteus* refused may be partly due to their being still coarser than the non-spined species, partly to the fact that the sticklebacks were always offered after the L. Even with this handicap, it is obvious from a comparison of these series with the foregoing ones that the protection enjoyed by sticklebacks is mainly due to their spines.

9. Discussion

It seems clear that both *Gasterosteus* and *Pygosteus* are, to perch and pike of the sizes used in these tests (10–25 cm) less attractive prey than four non-spined species of fish, and also, that the larger-spined *Gasterosteus* is much less vulnerable to such perch and pike than *Pygosteus*; the difference between *Pygosteus* and non-spined fish is, however, still demonstrable.

The series with the de-spined sticklebacks show that it is actually the spines that afford this protection. The sticklebacks' behaviour, which makes them raise their spines exactly when needed, and the structure of the spines and their erection mechanisms (studied by Hoogland, **8**) are both beautifully adapted to this function.

The rejection of sticklebacks by these predators is really based on two distinct processes. Many of the rejections were due to a non-conditioned response to the strong mechanical stimuli received after the predator had seized the stickleback. This response alone accounts for a sufficient number of rejections to prove the survival value of the spines, particularly since the overwhelming majority of sticklebacks survived after having been snapped up, and many of them survived even after very severe mauling. In addition to this, however, both perch and pike learn to respond to visual stimuli by avoiding sticklebacks before they have even touched them. All records scored as 'Avoidance' in Figs 142, 149 and 150 refer to this conditioned response.

The break in the reaction chain may occur either at the end of stalking, or already much earlier. Of the total of 162 observations recorded in series E (Fig. 151) in which the chain was interrupted before contact was made, 14 were broken off after the leap, 18 after stalking, 81 after turning the body towards the stickleback, and 49 already after the pike had merely turned its eyes towards the stickleback. It would certainly be interesting to know which the visual stimulus situation is to which the pike's avoidance becomes conditioned. From some preliminary tests it seems that the main characteristic is the way a stickleback moves; which of course is quite characteristic.

An ecological interpretation of our findings meets with some unexpected difficulties. We saw already that a starved pike, instead of learning to avoid sticklebacks, learns to eat them. We have no indications however that such a pike would actually prefer sticklebacks to non-spined fish, and in fact this seems unlikely.

A more serious problem is whether pike and perch of larger size would avoid sticklebacks. Our tests, all done with predators less than 25 cm long, naturally cannot answer this. However Frost (**5**) has made a thorough study of the stomach contents of pike of various size caught in Windermere, and her results seem in some respects contradictory to ours. Frost reports that minnows are much more abundant in the lake than *Gasterosteus*; that pike up to about 40 cm eat more minnows than *Gasterosteus*; and that larger pike eat more *Gasterosteus* than minnows. The percentage of minnows declines with increasing size of the pike; this may be explained by assuming that the optimum size of prey grows with increasing length of the pike. The percentage of *Gasterosteus* increases with increasing size of the pike up to the size-class 50–59 cm; this would mean that as the pike grows the spines bother it less. After that there is a drop, which may have to do with the shift in optimum prey-size. The puzzling feature is however that the percentage of *Gasterosteus* actually mounts above that of the minnows[1].

The only conclusion we can draw is that in Windermere *Gasterosteus* is on the average less well protected from pike than the minnow. In this connection it may be worth pointing out that, although we used minnows as one of four non-spined fish, our conclusion is not that sticklebacks are generally better protected than minnows, but that they are better protected than de-spined sticklebacks. The

[1] We must call attention to a misprint in Frost's paper which seems to contradict this statement. On p. 357 she mentions that the average numbers eaten are 0·16 for sticklebacks, 0·88 for minnows. From her table III to which this statement refers it is clear that the figure for the minnows is actually 0·08.

minnow may have anti-predator adaptations which work even better than spines, at least against large pike.

In this context it may also be worth speculating whether the conditions under which *Gasterosteus* live in Windermere are typical of the species. Since Frost (**5**) reports that 'sticklebacks, although at no time of great importance as food, are eaten mainly in May and June, at which time they congregate for breeding' (p. 351), her findings may mean that the stickleback's habit of migrating to shallow waters to breed has evolved as an adaptation which enables it to get away from large predators. Perhaps the Windermere population has too few shallow streams at its disposal and thus is subject to more severe predation in spring than other populations.

Whatever the solution may be, we believe that we are entitled to conclude that (1) sticklebacks derive a certain protection from their spines, and (2) *Gasterosteus* is much better protected than *Pygosteus*. The second point is worth elaborating. Morris (**9**) has shown that there are considerable differences in behaviour between the two species, and we suggest that our present study enables us to relate some of these differences with those in the effectiveness of the spines as anti-predator devices. *Pygosteus* is much more timid than *Gasterosteus*, and it is partly due to *Gasterosteus*' boldness that it is so well suited for laboratory studies of its behaviour. Any *Gasterosteus* brought into the laboratory in spring can be observed without special precautions within a day, even in tanks with very little cover. *Pygosteus* requires much cover, and even so remains shy for weeks on end, and often retires to the least accessible corners for nest building, so that mirrors, observation screens, etc., have often to be used in order to see its complete behaviour. It would seem as if *Gasterosteus* can afford to be 'tame' (and thus waste less time by fleeing) because of the protection offered by its spines.

Another difference between the two is in the type of nesting habitat they select. *Gasterosteus*, while preferring some vegetation, chooses a relatively open type of habitat; and often a few vertical weeds (offered in or outside the tank), or even crude visual imitation of weeds will induce it to settle and build. *Pygosteus* builds in dense weeds, and will only accept an open habitat after long habituation to the observer.

The difference in behaviour of the females is equally striking. Morris (**9**) has shown that female *Pygosteus* do not school in spring, but settle on territories of their own round those of the males, where they stay in cover most of the time. None of us has ever seen large schools of female *Pygosteus* roaming about in the spawning season, whereas the females of *Gasterosteus* are always easily seen, travelling about in schools up to 50 individuals strong, out of which the males pick those that respond to their courtship.

Finally, it seems not impossible that the difference in the nuptial colours of the males has something to do with the different need for procryptic coloration. Both species are usually camouflaged, showing general colour resemblance with the background, countershading, and disruptive crossbarring on the back. In the breeding season, *Gasterosteus* males turn a vivid red underneath; *Pygosteus* males become jet black. There seems little doubt that the red of *Gasterosteus* makes it much more conspicuous than the black *Pygosteus* males ever are. Moreover, during the actual sexual phase of the cycle, the back of *Gasterosteus* becomes a bluish white, which reverses the usual countershading, and makes the fish very conspicuous indeed, both when seen from above and when seen from the side. *Pygosteus* males also reverse their countershading, but the effect is far less spectacular.

All this leads us to conclude that *Pygosteus* has so to speak compensated for its inferiority as regards spine effectiveness by means of other antipredator devices such as better procryptic coloration, greater timidity, and a stronger tendency to keep in cover.

A number of authors have argued that *Gasterosteus* must have evolved from a *Pygosteus*-like ancestor. Bertin (**1**) gives anatomical evidence in support of this. A comparative study of nest building behaviour tends to support this conclusion, for *Gasterosteus* is the only species of stickleback that does not build its nest suspended in water weeds. In view of the evidence given in this paper it seems reasonable therefore to consider *Gasterosteus* as a species which has specialised in reduction of number and increase of size of spine, which has enabled it to reduce the time to be spent in fleeing and hiding in cover. This has resulted in various peculiarities of its reproductive behaviour, such as the selection of more open habitats, and the schooling and wandering behaviour of the females.

We believe that this is an example of the way in which diverse and seemingly independent systems such as anti-predator defence (involving morphological and physiological-cum-ethological components) and the reproductive system are in fact interrelated, and do not evolve quite independently. The two species, derived from common stock, have specialised in different types of anti-predator defence, and this has had repercussions in a wide field of other characteristics.

9. Summary

Feeding experiments with Pike (*Esox lucius* L.) and Perch (*Perca *fluviatilis* L.) as predators, and Three-spined sticklebacks (*Gasterosteus aculeatus* L.), Ten-spined sticklebacks (*Pygosteus pungitius* L.), Minnows (*Phoxinus phoxinus* (L.)), Roach (*Rutilus rutilus* (L.)), Rudd (*Scardinius erythrophthalmus* (L.)) and Crucian carp (*Carassius carassius* (L.)) as prey species show that both species of stickleback enjoy a demonstrable degree of protection from these predators. This protection is much better for *Gasterosteus* than for *Pygosteus*; tests with de-spined sticklebacks show that it is mainly due to the spines.

A description of the predatory behaviour of perch and pike is given; analysis of the experiments shows that (1) sticklebacks are rejected when, after being snapped up, their spines hurt the predator's mouth; (2) after very few experiences both perch and pike become negatively conditioned to the sight of sticklebacks and avoid them before they have made contact. As a result of this conditioning, the last links of the predators' feeding chain drop out; in pike, mere fixation with two eyes was often sufficient to recognise the prey as unwanted.

The possession of few and large spines in *Gasterosteus* must be regarded as more specialised than that of many small spines as found in various other species of stickleback. There is a correlation between the extreme development of this anti-predator device in *Gasterosteus* and (1) its boldness, (2) its tendency to select a more open nesting habitat, (3) the schooling and wandering tendencies of the females in the spawning season, and (4) its relatively conspicuous nuptial colours; these correlations suggest an interrelationship between anti-predator devices and reproductive behaviour and structures; the evolution of these two systems must have occurred in conjunction with each other.

REFERENCES

1 BERTIN, L. (1925). 'Recherches bionomiques, biométriques et systématiques sur les Epinoches (Gastérostéidés)', *Ann. Inst. Ocean. Monaco.*, **2,** 1–204.

2 BREDER, C. M. (1926). 'The locomotion of fishes', *Zoologica* (N.Y.), **4,** 159–297.

3 DEELDER, C. L. (1951). A contribution to the knowledge of the stunted growth of Perch (*Perca fluviatilis*), in Holland', *Hydrobiologica*, **3,** 357–78.

4 FRISCH, K. VON (1941) 'Über einen Schreckstoff der Fischhaut und seine biologische Bedeutung.', *Z. vergl. Physiol.*, **29,** 46–145.

5 FROST, W. E. (1954). 'The food of the Pike, *Esox lucius* L., in Windermere', *J. Anim. Ecol.*, **23,** 339–60.

6 HARTLEY, P. H. T. (1947). 'The natural history of some British fresh-water fishes', *Proc. Zool. Soc. Lond.*, **117,** 129–206.

7 HÖLLER, P. (1935). 'Funktionelle Analyse des Hechtschädels', *Morphol. Jahrb.*, **76,** 279–320.

8 HOOGLAND, R. D. (1951). 'On the fixing-mechanism in the spines of *Gastero-*

steus aculeatus L.', *Konink. Nederl. Akad. van Wetensch. Proc. series C*, **54**, 171–80.

9 MORRIS, D. (1954). 'The reproductive behaviour of the Ten-spined Stickleback (*Pygosteus pungitius* L.)', Doctor's thesis, Oxford University; also *Behaviour, Supplement* **6**.

10 WUNDER, W. (1927). 'Sinnesphysiologische Untersuchungen uber die Nahrungsaufnahme verschiedener Knochenfischarten', *Z. vergl. Physiol.*, **6**, 67–98.

Section IV
General Papers

Author's Notes

Ethology has often been called a new science. It will be clear by now that I think this is exaggerated. I prefer to consider it a re-emergence of an old science which, after periods of dormancy, of erratic and fragmented attempts at catching up with other sciences, and of hovering between the arts and the sciences, is now trying to make up for lost ground and to find its place among the modern life sciences—to become a branch of Biology in its widest sense.

When one has, as I have, taken part in such an exercise one has to 'sell' one's ideas and one's approach to others. The 'others' I have tried to reach have been first of all my fellow scientists. Not surprisingly, I have found many of them the least receptive of all; being engaged in their own pursuits, most of them have other bees in their bonnet than I have. A second group in my audiences consisted of students—those who aspire to become scientists and are still trying to make up their minds about the aims of Biology—they wonder what it is all about. I have always been happiest when dealing with them, but also most apprehensive, because no audience is as discerningly critical as students are. My third group of listeners was the interested general public. Like many of my fellow scientists I have always considered it part of my job to inform them about our activities; after all it's they who provide the public money from which our work is being financed.

While these three groups require different types of 'sales talk', I am convinced that these differences have often been exaggerated. In speaking to one's fellow scientists one can often do with a much smaller number of technical terms than is usual; and when speaking to the young and the non-scientist one need not, indeed must not, 'talk down'; to the contrary, most of them like to be taken seriously, and to be 'stretched'. I have therefore always tried to find a method of presentation that appeals to them all. But I am only too well aware of the difficulty of this task, and I am far from believing that I have found the best compromise.

Another difficulty is that, in trying to advance one's still adolescent science and at the same time to convince others of its value, there is a temptation to spend all one's time and energies either on pleading one's case in general terms, or on 'delivering the goods' in the form of concrete investigations, on the assumption that if the goods are found to be of sufficient quality the general message will in the long run come through. This problem of tactics, of finding the balance between these tasks, keeps occupying one's mind. Being at heart a missionary, I have always tried to do both to a certain extent. As far as teaching duties and informing the general public allowed—both demanding tasks—I have divided my time between either doing or (to an increasing extent) guiding concrete research, and salesmanship in the form of general addresses at congresses, symposia, etc., of which I have attended far more than I care to remember.

It was from the latter type of contributions that I had to select a few that I still judged worth reprinting, and this section contains those few. In drafting such general articles or addresses I was faced with the difficult choice of either trying to drive my points home by presenting an impressive array of facts taken from the literature as a whole, or taking selected examples from nearer home; examples taken from the work of myself and my own collaborators. For the sake of vividness of presentation (and as a result of poor reading habits) I have undoubtedly erred in leaning heavily on work with which I have been involved personally. That has led to a restriction in the range of studies quoted, and in consequence to a certain repetitiveness, even though the bibliographies of the papers quoted compensate for this to some extent. In retrospect this repetitiveness is a little embarrassing, but it has its advantages too: by slightly rephrasing one's arguments one 'clicks' now with this, now with that part of one's audience, and in order to contribute to a healthy public image of Ethology—today an over-fashionable, and much mispresented science—one has to keep searching for better forms of communication.

However, it has been precisely this awareness of the danger of over-repetitiveness (as well as the need of keeping the size of this book within reasonable bounds) which made me decide, reluctantly, to omit a number of papers—either on special species, such as sticklebacks, or on problems, such as displacement activities—which my colleagues urged me to include. An additional motive was my desire to present two contributions in which I touch on the relevance of animal studies to problems of human behaviour. Chapter 16 is a methodological discussion intended for a general audience, on a par with similar general treatises published by several of my colleagues, but the next article is an attempt to apply some ethological methods to a concrete and urgent problem of human psychopathology; it has elicited some surprisingly welcoming reactions, but also some violently rejecting comments—which I consider a sufficient reason for including it.

(*From the Department of Zoology of Oxford University*)

13
Behaviour and Natural Selection (1965)[1]

Introduction

We have learned from the preceding address by T. H. Bullock how complex the machinery is that controls behaviour, and that by W. H. Thorpe how numerous the interactions are between the genetically determined 'blueprint' and the environment that controls the development of behaviour in the individual. I now propose to turn to the question, 'What does an animal do with these intricate mechanisms? How do they help it survive and reproduce itself?' As a zoologist, confronted with the immense variety of animal types, I naturally will extend this question by comparing different species and applying the question in principle to any of them. Interest in this problem—considered by many to be on the fringe of biology—is reviving.

The view that characters, behavioural or structural, of which a function cannot be seen at a glance, might well be 'neutral', has in the past led to a more or less defeatist attitude. But the justification of such a hasty conclusion is now being seriously questioned (Cain, **6, 7, 8**; Pittendrigh, **37**), and the need for more systematic studies of survival value begins to be more widely felt. In this chapter I propose to discuss some lines of thought that have guided many recent field studies of animal behaviour. I have not tried to present an exhaustive review of results, and have drawn to a considerable extent on work done, or in progress in our own laboratory. I am indebted to my collaborators Dr E. Cullen, Dr H. Kruuk, and Dr I. J. Patterson for their permission to mention some aspects of their unpublished work.

In attacking our problem, let us start from observables—i.e. from behaviour. But instead of studying its causes we shall study its effects; in other words, rather than look back in time, as we do when studying

[1] Address delivered at a Plenary Session of the International Zoological Congress, Washington D.C., in July 1964.

causation, we investigate what happens as a consequence of the observed behaviour. I should like to stress that this *is* an investigation of cause-effect relationships, which as such requires experimental study as well as observation and speculation; it differs from the study of behaviour causation merely by the fact that the observed behaviour is the cause, and that its effects are studied; we follow events with time instead of tracing preceding events, and we determine an animal's success. What I shall ask is simply, 'Would the animal be less successful if it did not possess this behaviour?' or, a little more subtly, 'Would deviations from the observed norm be penalised, and if so, how?'

I shall approach this question as a naturalist—as one who delights in observing animals in their natural surroundings. Mayr has recently reminded us once more that 'the environment is one of the most important evolutionary factors' (**32**, p. 7). The naturalist knows perhaps better than any other zoologist how immensely complex are the relationships between an animal and its environment, how numerous and how severe are the pressures the environment exerts, the challenges the animal has to meet in order not merely to survive, but also to contribute substantially to future generations. He also realises how little we really know. Yet what we want to know is a great deal. We need to examine not merely *whether* a certain feature is of advantage to an animal; we also have to find out *how* it contributes to survival—that is, which requirement the feature meets (whether it assists in defence against predators, in feeding, in respiration, etc.); and further, how, (in terms of cause-effect relationships) it does so. Of course, such knowledge of the 'mechanism of survival' must ultimately be linked with the genetics of behaviour, so that the selection pressures against deviant genotypes in a given environment can be related to the genotypes that really occur.

In these few words I have tried to outline, in perhaps oversimplified terms, what is really a major programme of research, a programme which, in spite of much fascinating work already done, is still in great need of development. To quote Mayr once more: 'There are vast areas of modern biology, for instance . . . the study of behavior, in which the application of evolutionary principles is still in the most elementary stage' (**32**, p. 9).

The study of the way in which the characters of animals, whether structural or behavioural, contribute to survival has been discredited in the past by uncritical guessing where experimentation would have been required but for some reasons was not applied. As an example I remind you of Thayer's remarkably unrealistic picture (Thayer, **42**) of the way flamingos were supposed to disappear against a red sunset. Yet Thayer was also the man who, in the same work, pointed out the significance of 'countershading' as a component of camouflage employed by many animals. The correctness of this latter interpretation has since been experimentally demonstrated by de Ruiter (**41**). His work can be taken as an example showing (1) that survival value can be demonstrated as exactly as any other cause-effect relationship; and (2) that survival (in this as in so many other cases) is not due to a structural property alone (a colour pattern) but to the joint action of this structural character and specific behaviour patterns—in this case habitat selection, the adoption of the correct position, immobility, and scattering. It is, of course, this combination of structure and behaviour that makes the animal inconspicuous to visually hunting predators.

Camouflage, apparently a simple matter of matching coloration, is often the consequence of such complex co-ordination between structure and behaviour. As another example I mention the habit, found in many birds, of removing the empty egg shell after the chick has hatched. In the Black-headed gull it has been demonstrated (Tinbergen *et al.*, **49**) that failure to remove the egg shell allows certain predators to discover the otherwise camouflaged brood, and this seemingly trivial response has thus been shown to be a vital part of the defence against predators.

Studies such as these lead to conclusions of three types, which I shall emphasise because they have to be sharply distinguished.

1. They can provide actual, unequivocal proof of the way in which a feature contributes to survival or reproduction—in short, to success. They fill in the final links in our knowledge of the chain of events that determine survival; links that would not be discovered if one confined oneself to the study of causation of behaviour.

2. They can experimentally demonstrate, with equal force, selection pressures that prevent the species from deviating from its present state—in other words, they can provide proof of stabilising selection.

3. They can suggest (though no more than that) the selection pressures that have in the past moulded the species to what it is now.

This last interpretation rests on certain assumptions about the past that can admittedly be made very probable but which give this third conclusion a different status from the first two. When from now on I speak of selection pressures I do so in both senses (of stabilising and of moulding selection), and in full awareness of both this inherent uncertainty and the force of 'probabilistic' evidence.

Some Aspects of Method

In practice, this study has many ramifications and many methodological pitfalls. One can attack the problem from two sides.

1. The first method involves singling out a particular environmental pressure, a challenge to survival, and investigating how an animal meets this challenge. For instance, one can examine how a certain species defends itself against predators. My co-workers H. Kruuk (**24**) and I. J. Patterson (**34**) have been following this line of attack with the Black-headed gull, and have found that almost all its behaviour patterns have properties that make sense only as part of the total anti-predator defense system; we have seen that egg shell removal is part of this sytem; others are habitat selection, diurnal rhythm, colony density, and synchronisation of the breeding cycle.

The diurnal rhythm of habitat selection in early spring is as follows: All birds spend the night on their roosts, which are wide, open spaces such as the lower part of the seashore, open mud flats, or shallow lakes. Not until it is bright daylight do they occupy the breeding grounds, where they establish themselves on territories and proceed with the formation of breeding pairs. Until the eggs are laid the birds keep roosting on the open flats, but shortly after the start of laying they begin to sit through the night. This diurnal rhythm of habitat selection during early spring was supposed to have survival value long ago (Tinbergen **44**), the assumption being that wide flats are safer to the adults than the breeding grounds: on wide flats mammalian predators can be seen from afar and avoided, whereas on the breeding grounds, where the plant cover assists in concealing the camouflaged broods, this same cover might enable predators to kill the adults. Kruuk has now collected data on over 1,400 adult gulls killed by foxes and his data confirm that the gulls are practically safe on the beach (except on extremely dark, stormy nights) but are preyed upon to a considerable extent on the breeding grounds.

Data obtained by Patterson and Kruuk also begin to show the adaptedness of the particular density pattern of gull nests (Fig. 152). On the one hand, social nesting has been shown to allow the gulls to attack in force bird predators such as crows and large gulls; such mass attacks are highly effective (**24**). This, and the ineffectiveness of attacks by few gulls, put a premium on dense nesting in this species. On the other hand, there is suggestive, though as yet no conclusive evidence indicating that too dense nesting might endanger broods with respect to those predators that are not repelled by attacks, e.g. foxes. The particular density of a gull colony (or more precisely, the range of densities, for the density varies with the habitat) seems to be the result of at least these two opposing pressures, both exerted by predators. Although other factors may be involved as well, my impression is that predation pressure is overriding.

Patterson (**34**) finally has shown that laying is well synchronised;

most clutches are laid in a peak period of approximately fourteen days. By comparing mortality of early, 'peak' and late broods, he found that early and late broods are penalised much more severely than are peak broods—in fact, hardly any late broods survive at all (Fig. 153).

The behaviour mechanisms responsible for these three adaptations are only very superficially known.

Or—to take another type of selection pressure—one can study the differences between closely related species or subspecies, one of which has to coexist sympatrically with another form while the other has

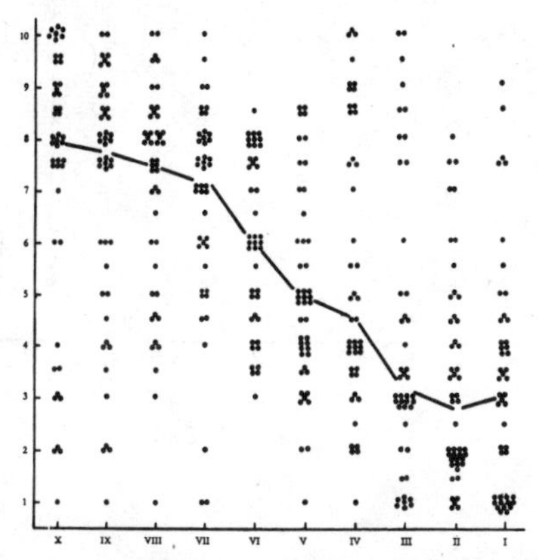

Fig. 152. Order in which the eggs were taken from the egg-lines by crows and Herring gulls. Horizontal is egg position (X is inside, I is outside the colony), vertical is the order of disappearing of the eggs (e.g. the bottom right-hand cluster shows that in 1.3 tests the egg in position I was taken first, etc.). The solid line connects median preferences.

not. As one example out of several now known I select W. F. Blair's work (**3**) on *Microhyla olivacea* and *M. carolinensis*, in which he found that, where the two species overlap, their songs are more different than where they do not coexist. In view of what we know about the function of Anuran songs as signals attracting females it is clear that, in the area of overlap, selection pressure for interspecific diversity of the songs has been at work, which has resulted in more perfect sexual isolation. Perdeck (**35**) has shown a similar function of song-specificity in two sibling species of grasshoppers; there is little doubt that the same phenomenon is found in the song patterns of other birds (Marler, **31**); many birds and fish have developed visual signals subserving sexual isolation (Tinbergen, **47**), and it has recently become clear that chemical signalling shows the same diversity wherever there is need for unambiguity (Wilson, **52**, **53**). Here the selection pressure on which the work centres is not predation—it is the disadvantages of hydridisation.

The effects of specific selection pressures can also be studied by direct application of a pressure and study of its evolutionary effects.

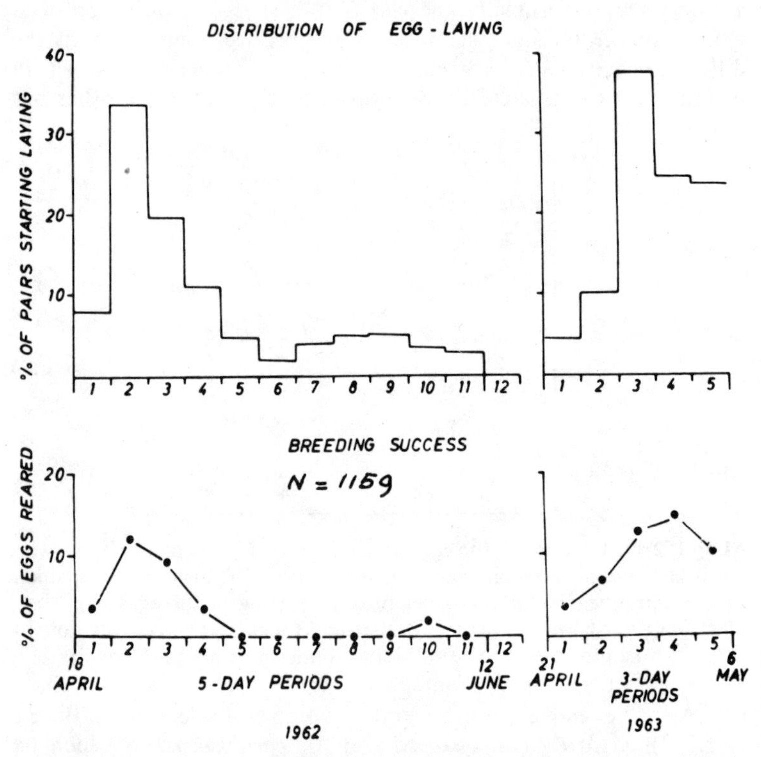

Fig. 153. Changes in breeding success of Black-headed Gulls with laying date, compared with the distribution of laying. The number of eggs on which each percentage is calculated is written beside the point. The histograms are based on 406 nests (1962) and 208 nests (1963).

For instance (to mention another example concerning sexual isolation), Crossley (**11**) applied 100 per cent antihybrid selection in mixed populations of two strains of *Drosophila melanogaster*, and found that, in forty generations, the females of both strains had become more selective in their responses to males of the two strains, while males had become more sensitive to the repelling behaviour of strange females—both trends, of course, contributing to the increased sexual isolation that she actually found at the end of this experiment.

2. The second method starts from the other end, from an observable property of an animal of which the function is unknown, and examines in what respect it contributes to survival—which requirement imposed by the environment it meets, and how it does so. For instance, when one observes that Black-headed gulls synchronise their breeding cycles more closely than do most other birds, one can ask whether this has a function. We have seen that one of my collaborators, I. J. Patterson, has compared the success of large numbers of 'peak broods' with that of later and of earlier broods, and he finds that peak broods are more successful than either. He also determined, in collaboration with H. Kruuk, the environmental pressure responsible for this: predation was higher for the late broods, and probably higher for the earlier broods as well, than it was for the peak broods. The so-called fanning of the male Three-spined stickleback may serve as another example (Tinbergen, **43**; van Iersel, **20**). Fanning is done mainly when there are developing eggs in the 'nest'; it involves a particular type of coordination of the movements of the pectoral fins and the tail, which results in water being driven through and over the nest entrance. When fanning is prevented, or when it is allowed but the nest is covered with a watch glass, the eggs die. They develop normally in the absence of a male, provided a freshly aerated water current is led through the nest. These simple tests show that the fanning serves to ventilate the eggs, and in this context it is adaptive that the rate of fanning is controlled, from without and from within, in such a way that ventilation is always adequate.

Gwinner and Kneutgen (**15**) discovered the function of a hitherto puzzling feature of the vocalisation of some song birds. Many species learn their song by listening, often during a short critical period, to fellow members of the same species; some even mimic many different sounds. The Schama thrush, *Copsychus malabaricus*, is one of the more elaborate mimics, and different individual males have quite different and therefore individually distinct song repertoires. Usually, the female of this species does not sing, but when the partners (the species is monogamic) become separated, she at once begins to utter all the individually acquired phrases typical of her own mate, which she must have learned by 'latent learning'. The male responds by replying and returning to her. The same effect can be obtained by playing a tape recording of the male's voice back to him when he has drifted from his territory; under these circumstances he attacks the loudspeaker. The female, so to speak, parasitises on the male's territorial aggressiveness to get him back.

The main value of these studies lies, I suggest, in the inspiration they provide, in the concrete problems they bring to light that have to be solved if we are ever to proceed beyond merely proving the fact that selection is a powerful evolutionary agent—if we are to understand to what extent selection has been responsible for making the present animal species into what they are. And this I consider to be the crucial point of my plea: what the selectionist has to do is not to demonstrate the fact 'that the fittest survive'—that phrase, if applied to a heterogeneous population, is a tautology anyway—nor merely to show differential survival or success, but to test whether the properties of organisms really do contribute to their fitness, and if so, how they do so.

I should also like to stress, however, that we can do more than receive inspiration: we can, with some tact and a great deal of good luck, collect a considerable amount of experimental evidence.

The inspiration, which leads to hypotheses, can be derived in various ways. I should like to mention three sources that have proved particularly fertile.

1. Animals kept in captivity or semi-captivity—which means in an abnormal environment—often show behaviour that, under the circumstances, makes no sense—it 'misfires'. The reaction of the functional biologist is, 'This must have a function—let us see what happens when the animal does it in its natural habitat'. Thus Lorenz (**27**, **28**) found that his captive jackdaws would repeatedly fly low over his head and wag their tails just in front of him. This made him look at this behaviour in wild jackdaws, and he discovered that it is a signal by which jackdaws stimulate members of the flock to join them in flight. Another example: young ducklings and birds of several other species perform peculiar trampling movements when they are placed on a wet substrate (Tinbergen, **46**). In the natural environment this is part of the feeding behaviour: in shallow ponds the paddling stirs up motionless or concealed animals, which are then seen and eaten. Once one has discovered this function, it becomes a matter of interest that the paddling is combined with looking down. Incidentally, M. Rothschild (**38**) reports that one of her hand-raised Black-headed gull chicks, which was born blind, made the same 'looking down' head movement when paddling as normal chicks do.

Another example of misfiring is provided by the observation I did years ago, when I was conducting a practical course on animal behaviour at Leiden University. A row of tanks, each containing a male Three-spined stickleback in reproductive condition, was set up each spring in front of large windows overlooking a main thoroughfare of the town. Whenever a red Royal Mail van drove past, all the males would dash at the window side of their tanks, keeping their heads oriented at the moving van and making frantic attempts to swim through the glass. This gave rise to experiments, summarised in Tinbergen (**43**), demonstrating that the red underside of male *Gasterosteus* acts as an epigamic signal.

A Peacock butterfly (*Vanessa io*) has cryptically coloured ventral wing surfaces. When at rest, these surfaces are exposed to view. When disturbed in cool weather, for instance by a human observer prodding it with a sharp twig, the Peacock butterfly flaps its wings, thereby exposing the brightly coloured 'eyespots' on the dorsal surfaces and its fore and hind wings. While doing this the insect orients itself accurately in such a way that the surface of the wing is continuously turned towards the observer. Observations such as these led to experiments by Blest (**4**) that demonstrated that such eyespots scare off insectivorous song birds.

2. Closely related species may differ strikingly in their behaviour. The functional biologist, intent to know why each of these species behaves the way it does, pays equal attention to behaviours a species possesses as to those a species lacks; he also wants to understand why two roughly similar behaviours differ slightly between species.

The two European freshwater sticklebacks, *Gasterosteus aculeatus* and *Pygosteus pungitius* differ consistently in the threshold of their fleeing responses, *Gasterosteus* being the tamer of the two (which made it such a suitable object for laboratory behaviour studies). With my co-workers R. Hoogland and D. Morris (Hoogland, Morris, and Tinbergen, **19**) I found that this was correlated with the difference in spine armature of the two species. Both species could be shown to derive protection against certain predators from their ability to erect and lock their spines when caught; the long, strong spines of *Gasterosteus*, however, are far more effective than the small spines of *Pygosteus*. Correlated with this is not merely the shyness of *Pygosteus* and its habit of dashing into cover when threatened, but also the selection of an open habitat by *Gasterosteus* and of a densely overgrown habitat by *Pygosteus*; further, *Gasterosteus* has been able to afford bright nuptial colours (including a striking reversal of the normal countershading), whereas the nuptial colour of *Pygosteus* is an overall pitch black with, as the only concession to the demand for conspicuousness, two brightly silverish-green ventral spines, which are displayed only during the courtship dance, and even then can be seen only from behind—which is where the female is.

The comparative method has also been applied in E. Cullen's (**12**) study of the Kittiwake (a pelagic gull); she shows convincingly that numerous peculiarities of this species are adaptations to its habit of nesting on narrow ledges on sheer cliffs—itself an antipredator measure. Crook (**10**) has recently shown in his extensive studies of the behaviour of Weaver birds how extremely fertile this approach is.

3. Comparison of non-allied species, particularly of those that, while very different in general, show some striking similarities, is equally fruitful as a method, because the similarities are, of course, suspected to be the result of convergent adaptation. A fine study applying this method is the paper by von Haartman (**17**) on the adaptations, behavioural and otherwise, of birds nesting in holes—which itself, like the cliff nesting of many seabirds, affords protection from predators.

All camouflaged animals share some behavioural adaptations that have not sufficiently been emphasised, nor studied in detail. They usually stay motionless as long as there is the possibility of their being seen, i.e. during the daytime—and move only when there is an overriding need. It is only mobile species well equipped with sense organs that can afford to move about, though even these usually keep in cover and rely on motionlessness when a predator is in sight. Camouflaged species further select a habitat that matches their colour, or they have means of adapting their colour to the background. Third, they tend to select a position in which their crypsis is most effective; for instance, countershaded animals always rest in a position in which their darker side is uppermost. Finally, cryptic animals live scattered. This has undoubtedly to do with the fact that, while a predator can find cryptic prey, it has to search for them. It is even probable that this searching is controlled by a temporary narrowing of the attention (the formation of a 'searching image') and that the persistence of such a searching image in the predator, and hence its effectiveness as a mortality factor of its prey, is dependent on the rate of being rewarded. This rate is, of course, kept low when the prey lives scattered.

Elsewhere (Tinbergen, **45**) I have listed a few of the many different mechanisms used by various animal species for the purpose of spacing-out—whether as part of their camouflage or for other reasons. From that brief review it was obvious that further studies of dispersion mechanisms are badly needed.

Only part of the studies mentioned contain experimental confirmation of the functions suggested. Yet I should like to emphasise their great value. It cannot be stressed too much in this age of respect for—one might almost say adoration of—the experiment, that critical, precise, and systematic observation is a valuable and indispensable scientific procedure, which we cannot afford to neglect (see Lorenz, **29**). Particularly in our young science we need good observers, and the sense of hurry, the urge towards spectacular 'breakthroughs' must not be allowed to lead to a kind of contempt for nonexperimental observation, which admittedly is a slow procedure but which, by a trial-and-error process, has to provide us with our 'hunches'.

Although it is, of course, true that the evidence discussed so far does not amount to hard and fast proof, research often stops at this point. This is due to a variety of circumstances. First, some workers consider this kind of evidence convincing enough in itself. To take a crude example: one need not cut off a blackbird's bill to prove that it assists in feeding, nor to observe the failure of a butterfly that has lost a wing to be convinced that wings are indispensable for flying. But another reason for not proceeding to experiment is the fact that it may be difficult or almost impossible in practice. With some *structures* experimentation is easy. For instance, in order to show that the so-called eyespots on the wings of Saturniid moths have the function of scaring off song birds, Blest (**4**) compared the effect on song birds of normal moths, and of moths whose eyespots had been brushed off, thus comparing, as a good experiment should, two classes of animals that differed in one special, known aspect and in nothing else. Such a comparison of the success of two forms, one possessing a certain character, the other lacking it while being otherwise identical, is, of course, much more difficult in behaviour studies, for one cannot usually 'dissect' just one behaviour pattern. In simple cases this can be done: for instance, one can, by a simple operation, silence male locusts without affecting any other behaviour, and thus show the function of stridulation as a sexual attractant (Busnel, Dumortier, and Busnel, **5**). Or one can play back sound recordings and thus present the sound in isolation. These are only special cases of a method widely used in ethology: the use of dummies in which single characters can be changed at will. Even where such isolation is not possible, one can proceed in a roundabout way. For instance, the fanning of a male stickleback has been shown to be indispensable for the ventilation of the eggs by a combination of experiments that together leave no room for doubt (van Iersel, **20**). So experiments on survival value can be conducted in a variety of ways.

However, this first experimental step is but a beginning. Most of these experiments are alike in having been done in 'controlled' situations. This means, however, that they have been done in impoverished environments, and this in turn means that only one or a few effects, rather than all possible effects, have been studied. Yet it is the overall effect in the natural situation by which we must judge survival value. The experiments mentioned above that showed that failure to remove the egg shell exposed the brood of gulls to higher predation did not take into account that the parent gull, in removing the egg shell, leaves the brood unguarded for some seconds. These few seconds can, however, be sufficient for some predators to snatch quickly an egg or a chick. The overall value of egg shell removal is the resultant of this disadvantage and the advantage shown up in the experiments. As we have seen, Kruuk has shown, also in the Black-headed gull, that nesting in dense colonies is of advantage in allowing mass attacks on crows and Herring gulls, which actually repel these predators only when the attack involves sufficient numbers. Yet again this mass attack makes the birds leave their nests, when a determined predator can break through and grab an egg or chick. The total effect of social attack is the resultant of these, and perhaps of more effects—and this has to be measured in the natural environment. Studies in which this overall effect has been studied, as well as the separate pressures and agents involved, are still extremely rare. A fine example can be found in Kettlewell's work (**21**, **22**) on the part played by song birds in controlling the evolution of camouflage in the Peppered moth. His study has the additional merit that the genetic difference between the two forms was known.

The work discussed so far is incomplete in another respect. I have been considering absence and presence of a character, or at least very large differences. However, we are not merely interested in knowing whether or not the total absence of a character is penalised: we want to find whether the character as found now is optimal or not, whether slight deviations from the norm are penalised. Are slightly less 'perfect' eyespots of moths less effective? Are slightly less twig-like caterpillars less successful? Are female kittiwakes that are slightly more timid than normal at a disadvantage? In these three cases we happen to have evidence indicating that small deviations are indeed penalised (Blest, **4**; de Ruiter, **40**; E. Cullen, pers. comm.). When we realise how powerful a selective advantage must be before we can show it up in our crude and small-scale experiments, it is amazing that it has been possible to show an advantage at all. It is well to keep this in mind in view of Fisher's (**16**) and Haldane's (**18**) calculations about the minute selective advantages that can be sufficient for selection to produce evolutionary change.

However, in other cases it has been shown that the norm is not optimal with respect to certain specific requirements. The best-known examples are found among the social signals. In 1951 (**43**) I listed the cases in which it had been shown that dummies could be made 'supernormal'—that is, more stimulating than the natural object. Weidmann (**51**) has since shown that the red bill of an adult Black-headed gull, to which the chick responds when begging for food (which in turn stimulates the parent to feed it) elicits fewer responses in the chick than a black cardboard 'bill' with a contrasting white dot. Such facts lead, of course, at once to the question 'why has not selection produced a better bill?' No answer to this can be found unless one studies the animal's entire behaviour repertoire and the animal's complete natural environment—I have to emphasise that a preoccupation with the need of experiments in controlled conditions would *prevent* us from finding the solution. One answer, often proposed but rarely fully documented experimentally, is found in the fact that the many different selection pressures acting on an animal are often contradictory or conflicting, and that selection (which rewards or penalises the phenotype, not isolated characters) has produced a compromise. As I hope to discuss later, concrete knowledge of these compromises is indispensable in order to meet certain objections against applications of the selection theory that are based on lack of knowledge, in fact on mere incredulity rather than on positive knowledge.

For an example of such conflicting pressures I return once more to eggshell removal in the Black-headed gull. This response shows a characteristic that seems at first glance to be suboptimal: the shell is not removed at once when it breaks, but on the average not until one to two hours after hatching (Tinbergen *et al.*, **49**). One counter-pressure preventing promptness is that the chick needs time to free itself completely from the shell lest it should be removed with the shell. This is safeguarded by the parent stopping the response when, upon seizing the rim of the shell in its bill, it feels the weight of the chick (Tinbergen, Kruuk, Paillette, and Stamm, **50**). Another counter-pressure is exerted by neighbouring gulls, some of whom prey selectively on pipped eggs and newly hatched, still-wet chicks (whereas dried, fluffy chicks, being difficult to swallow, do not interest them). As a rule the parent does not remove the shell until the chick has passed this dangerous stage. Several of these conflicting demands exists in the antipredator system of this gull; they have been analysed and listed elsewhere (Tinbergen *et al.*, **49**).

To return now to the colour of the gulls' bills: it is probable that this problem requires a similar explanation. The most likely hypothesis, as yet unproven, is that the red bill of the Black-headed gull is an epigamic signal, or is effective as a threat in territorial spacing-out.

It is becoming increasingly clear that these interactions of selection pressures are of immense complexity, and also that, without detailed functional study, one cannot even begin to understand adaptedness. E. Cullen's study of the Kittiwake, already mentioned, shows convincingly that such peculiarities as the building of a mud platform under the nest, the immobility of chicks, the method of fighting by 'bill-twisting', the absence of one of the threat postures generally found in gulls, the habit of guarding the nest even before eggs are laid, and numerous other characters are adaptive, and are all corollaries of one antipredator device: nesting on sheer cliffs. Similarly, Crook shows in his penetrating comparative analysis of the behaviour of Weaver birds that the kind of niche a species occupies affects many details of their mating behaviour irrespective of affinity. We are only just beginning to scratch the surface of these problems.

Results

These few examples will, I hope, be sufficient to show the magnitude of the field we have to explore before we can hope to assess to what extent and in what respects the behaviour of animals is adapted to the needs imposed by the environment. I have argued the need of exploring this field more fully, and I have claimed that perfectly sound methods are possible and are, in fact, being applied, though not as yet very systematically. What can one say, in general terms, about the results so far achieved? I should like to draw attention to two points.

1. *The search for the mechanism of survival value has to be made with respect to all the properties of behaviour mechanisms*, all the details of behaviour control. The most obvious aspect of behavioural adaptedness is, of course, the specialised, 'highly improbable' form of the motor patterns themselves. Various birds, for instance, have specialised movements that 'take the sting' out of a bee (Nicholls and Rook, **33**); the intricacies of adaptation of the sounds used for echolocation by bats are being revealed in all their astonishing detail (Griffin, **14**). Many data on adaptedness are now available on signals serving communication, where structures (sound- or scent-producing organs, colour patterns) have been developed jointly with the behaviour employing them (Lorenz, **28**; Lanyon and Tavolga, **25**; Marler, **31**; Tinbergen, **47**). But adaptedness of movement is less well known, and is yet not less striking in simpler activities, such as the way a thrush captures an earthworm, or the way a Godwit walks through tall grass; this it does by lifting and folding its feet—which a Lapwing, adapted to habitats with low vegetation, cannot do. Klomp (**23**) has shown that this seemingly trivial peculiarity of the Godwit has survival value.

Apart from the motor pattern employed, their sensory control is delicately adapted to the needs. The problem of selective responsiveness applies to stimuli characterising the appropriate class of objects to which the response makes sense—de-stinging is done only with wasps and bees; how are these insects recognised? Signals (to mention but one example) must be used only when needed, for their very conspicuousness invites predator attack. Further, highly adaptive orientation mechanisms have the function of ensuring proper relationships in space. There must also be appropriate control of the quantity or intensity of the response: for a proper functioning of territorial spacing-out an animal must not be too easily aroused to attack, nor must it be too timid (Tinbergen, **45**). It is useful in species that rely on fast escape to give stimuli from predators precedence over those given by food; yet species that protect themselves from predators by selection of a safe habitat must not be too ready to take flight, lest they should waste time and energy that can be usefully employed in other ways. Escape from predators may have to be dependent on the part of the habitat an animal happens to be in: the marine Galapagos lizards are fearless on land, but shy under water, where they may meet predators (Eibl-Eibesfeldt, **13**); Kittiwakes are fearless on the cliff, but shy when collecting nest material on land (Cullen, **12**). Some behaviour patterns such as distraction displays can be allowed to wane easily since the occasion for their employment arises rarely (Lorenz, **28**), but other responses, such as the searching behaviour that preceded the catching of prey, must not wane (Precht and Freitag, **38**).

Finally, even the particular type of ontogenetic control is itself part of the adaptive equipment of a species. Where a behaviour pattern must function nearly perfectly the very first time it is performed (such as flying in a newly hatched butterfly or a fledgling Gannet), a largely internal control of development, or 'innateness', has survival value. But in other cases it is clearly of advantage not to be too precisely adapted innately but to leave a great deal to trial-and-error learning. For instance, many young birds try to eat a wide variety of objects, edible and inedible alike, and have to learn what is edible. This enables them to specialise individually on the type of food most readily available in the particular habitat in which they happen to grow up.

Even where a great deal of learning determines the characteristic behaviour patterns of a species, the ability to learn is often not a general property but is laid down in 'innate dispositions to learn'. Honeybees have the tendency to remember the location of an abundant food source, which many other insects have to a much lesser degree. Baerends (**1**) has demonstrated elegantly that a solitary wasp, *Ammophila adriaansei*, has a truly remarkable capacity of delaying a response at one special stage in its reproductive cycle: stimuli received during one single 'inspection visit' to one of its burrows can determine for many hours how many caterpillars the wasp captures and where she takes them—an achievement of which many vertebrates could be proud.

Of particular interest are those responses having the function of creating an opportunity to learn. Exploratory behaviour is one of these, and the 'locality studies' of bees and wasps have been shown to have as their main and probably only function the creation of an opportunity to learn landmarks (Tinbergen and Kruyt, **48**). The full story of their adaptedness is certainly not yet known. Manning (**30**) found that bumblebees make locality studies when leaving newly discovered *Cynoglossum* plants, and that they thus learn the location of individual plants, which is highly adaptive since the small flowers of *Cynoglossum* are visible from a short distance only; but no locality studies are made when the bees leave Foxglove plants (*Digitalis*), whose large flowers can be seen from a great distance. It obviously 'does not pay' to learn the position of individual *Digitalis* plants.

These few facts, selected almost randomly, may be sufficient to illustrate my first point: the need for studying the survival value of all aspects of behaviour-control mechanisms.

2. My second point, which is closely related to the first, is that *the facts just mentioned show how useless and even unscientific it is to pronounce an opinion on the question whether or not 'a certain feature is adaptive' until one has tried to find out.* Yet adaptedness is often denied on occasions when, in the evolution literature, the question is discussed to what extent the properties of present-day animals can have been produced by natural selection. Let me therefore, in conclusion, review briefly what investigations of the type I have been discussing can contribute to this question.

Conclusion

Even in the minds of those who consider that natural selection as a principle is firmly established, one still finds doubts about, and widely conflicting assessments of its relative importance. These doubts are of various kinds.

1. The obvious imperfection of many characters leads to the question, 'If selection were so important, why is not the animal better equipped?' We have seen that functional studies, which take the total environment into account, show that the need for compromise is one important reason.

2. Paradoxically, some characters seem to be so amazingly perfect that one wonders whether a selection pressure with such exacting demands actually exists. For instance, the 'eyespots' of some moths

and the twig mimicry of some insects seem more perfect than can possibly be required—it has been argued that not even a bird could possibly detect small deviations. The only way to find out is, of course, to try. And actually Blest found, with respect to eyespots, as did de Ruiter with respect to twig mimicry, that surprisingly small defects were detected by the birds that act as selectors—in fact some birds are much more critical selectors than we would be.

3. It is often argued that some characters that now have survival value 'could not possibly' have had survival value when they first appeared in their crude, incipient form. This argument too has been applied to eyespots. But Blest, who took the trouble to experiment, found that even very simple contrasting spots had some effect on song birds.

4. It is often said that some forms of social cooperation and interaction 'could not possibly' have arisen through selection on the level of the individual or of the breeding pair, and must therefore be due to group selection. This argument has recently been revived by Wynne-Edwards (**54**) whose book is concerned with the ways in which animals prevent overpopulation. It contains two main theses: first, many animals have developed means, usually behavioural, of preventing overcrowding; second, many of these means are 'altruistic'—that is, beneficial to the population as a whole but not to the individuals—and as such can only be explained as consequences of group selection.

While I believe Wynne-Edwards's first thesis to be sound—even though he seems to apply it to many phenomena that may well have other functions—his second thesis has the weakness of being based on negative evidence, on lack of analytical data. The argument hinges on the hypothesis that certain forms of social cooperation are really 'altruistic' in the sense that they are against the interest of individuals. Now, if it could be shown by concrete analysis that such forms of social interaction could arise as the result of conventional natural selection, such a theory, though of course not disproved in principle, would lose the only type of support that Wynne-Edwards marshalls in its favour. I can mention two examples in which such analyses show that seemingly altruistic behaviour can be the result of the joint action of two selection pressures, each acting on the level of the individual or the breeding pair.

Many seabirds fish socially, each bird being attracted by the sight and perhaps sound of an individual that has discovered accessible fish. For gulls it has been argued that the white colour has been developed as an 'altruistic' character attracting flock members to food. My collaborator, Phillips (**36**), showed experimentally that the white coloration of gulls and terns has survival value to individuals as 'aggressive camouflage': it allows them, to put it simply, to approach fish unseen. He argues that, as long as this advantage overrides the disadvantage (if any) of being joined by others, a species would of necessity develop social fishing if it experienced a second selection pressure, also acting on the level of the individual: a pressure promoting selective *responsiveness* to a successfully fishing white bird.

Wynne-Edwards also calls upon group selection to explain the existence of territorialism as a means of spacing out. But as I have argued elsewhere (Tinbergen, **45**) there are many good reasons to conclude that territorial spacing-out is effected by a balance between two behaviour systems, each of which is of advantage to the individual and its own brood: that of attacking rival males when one is on one's own territory, and that of fleeing from a rival male when one is not on one's territory. There is nothing altruistic about territorialism, even though all members of the community profit.

Whether or not an evolutionary mechanism for group selection in Wynne-Edwards's sense is possible is not at issue; the point at issue here is that Wynne-Edwards' conclusion is based on absence of evidence; in fact, as Cain (**8**) has correctly phrased it, on *incredulity*. Whatever the truth about group selection may be, it must not be allowed to provide a way out of a problem that can be tackled in a positive way.

5. A somewhat similar argument applies to cases in which genetic drift is invoked to explain characters assumed to be nonadaptive. It seems to me an abuse of an important principle to apply it, so to speak, *in vacuo*, namely, to features called nonadaptive before even the attempt is made to investigate their possible survival value. The work of Cain and Sheppard (**9**); see also Cain (**8**) provides a striking example of the usefulness of a character first thought of as nonadaptive.

6. It is often said, when the adaptive value of a character is not at once obvious, that 'it could well be due to pleiotropism'. This is a case of false alternatives: pleiotropism refers to a type of growth control in the individual, whereas selection applies to an evolutionary process, which produces types of growth control, and which has a time scale of a quite different order. Selection may, so to speak, have utilised or manipulated pleiotropic growth mechanisms so as to produce advantageous results. When the relations of an animal with all aspects of its environment are studied, numerous interactions between selection pressures are discovered that have led to compromises, and this fact should encourage us to look for indirect

selection effects. Without studying why it might be necessary for a Black-headed gull to delay the removal of the egg shell until the chick has dried, one cannot possibly say whether or not this lack of promptness, this particular timing mechanism, is adaptive. To shrug off the need for a functional study by saying that such a character could well be due to pleiotropism amounts to a refusal to investigate.

These six issues, which I have touched only briefly, have one thing in common: in all of them one finds, even in recent literature, a tendency to base a positive judgment on absence of information. Yet, as I hope to have shown, there are ways of acquiring relevant information.

One could, of course, argue—as a distinguished physiologist recently did in a private conversation—that all this is flogging a dead horse, since no biologist has any doubts about the correctness of the selection theory. I think it is worth stressing that to accept this as an excuse for not investigating survival mechanisms *in concreto* is comparable to resting content with the general statement that life processes are ultimately based on molecular processes, and refraining from concrete biochemical and biophysical analysis.

Summary

It has been my purpose to argue, with a few selected examples, that:

1. Perfectly good methods are available for the study of environmental pressures imposing demands on an animal's behaviour.

2. For a proper evaluation of these pressures the animal must be studied in its natural environment.

3. If this is done, a picture of surprisingly intricate adaptedness emerges. And, finally,

4. It then becomes clear that we can show up certain weaknesses in evolutionary thinking that are based on incredulity rather than on evidence.

In conclusion I should like to refer once more to Mayr's statement, quoted at the beginning, to the effect that evolutionary studies of behaviour are still in the most elementary stage. It will be clear that I agree with this statement on the whole, and I would not quarrel with those who would say that the rather pedestrian level of my talk bears out Mayr's opinion. Yet I believe that in one respect behaviour studies may well be ahead of other fields of evolution study. Many ethologists are by tradition field observers, and they are thus in a good position to see the animal in its continuous struggle with the very complex environment that has moulded them. I believe that the type of work I have been discussing might well lead to a better knowledge of the powers of natural selection in general, and to a higher regard for these powers with respect to all functions and structures as we find them in present-day animals.

REFERENCES

1 BAERENDS, G. P. (1941). 'Fortpflanzungsverhalten und Orientierung der Grabwespe *Ammophila campestris*' Jurr, *Tijdschr. Entomol.*, **84,** 68–275.

2 —— (1950). 'Specializations in organs and movements with a releasing function,' S.E.B. Symp., **4,** 337–60.

3 BLAIR, W. F. (1955). 'Mating-call and stage of speciation in the *Microhyla olivacea—M. carolinensis* complex', *Evolution*, **9,** 469–80.

4 BLEST, A. D. (1957). 'The function of eyespot patterns in the Lepidoptera', *Behaviour*, **11,** 209–56.

5 BUSNEL, R. G., B. DUMORTIER, and M. C. BUSNEL (1956). 'Recherches sur le comportement acoustique des Ephippigères (Orthoptères, Tettigoniidae)', *Bull. biol. Fr. Bels.*, **90,** 219–86.

6 CAIN, A. J. (1951). 'So-called non-adaptive or neutral characters in evolution', *Nature, Lond.*, **168,** 424.

7 —— (1951). 'Non-adaptive or neutral characters in evolution', *Nature, Lond.*, **168,** 1049.

8 —— (1964). 'The perfection of animals', in *Viewpoints in Biology*, J. Carthy, and R. Duddington (eds.). vol. 3, pp. 36–62, Butterworth, London.

9 —— and P. M. SHEPPARD. (1954). 'Natural selection in Cepaea,' *Genetics*, **39,** 89–116.

10 CROOK, J. (1964). 'The evolution of social organisation and visual communication in the weaver birds (Ploceinae)', *Behaviour*, *Suppl.* **10**.

11 CROSSLEY, S. A. (1960). 'An experimental study of sexual isolation within the species *Drosophila melanogaster*', *Anim. Behav.*, **8,** 232–3. (Full account in unpublished Doctor's thesis, Oxford.)

12 CULLEN, E. (1957). 'Adaptations in the kittiwake to cliff nesting', *Ibis*, **99,** 275–302.

13 EIBL-EIBESFELDT, I. (1962). *Galapagos—die Arche Noah im Pazifik*, München.

14 GRIFFIN, D. R. (1958). *Listening in the Dark*. Yale University Press, New Haven.

15 GWINNER, E. and J. KNEUTGEN (1962). 'Ueber die biologische Bedeutung der "zweckdienlichen" Anwendung erlernter Laute bei Vögeln', *Z. Tierpsychol.*, **19,** 692–6.

16 FISHER, R. A. (1930). *The Genetical Theory of Natural Selection*. Dover Publications, New York.

17 HAARTMAN L. VON (1957). 'Adaptations in hole-nesting birds', *Evolution*, **11,** 339–48.

18 HALDANE, J. B. S. (1932). *The Causes of Evolution*. Longmans Green, London.

19 HOOGLAND, R., D. MORRIS and N. TINBERGEN (1957). 'The spines of sticklebacks (*Gasterosteus* and *Pygosteus*) as means of defence against predators (*Perca* and *Esox*)', *Behaviour*, **10,** 205–36 and this volume p. 52.

20 IERSEL, J. J. A. VAN (1953). 'An analysis of the parental behaviour of the male Three-spined stickleback', *Behaviour*, *Suppl.*, **3,** 1–153.

21 KETTLEWELL, H. B. D. (1955). 'Selection experiments on industrial melanism in the Lepidoptera', *Heredity*, **9,** 323–42.

22 —— (1956). 'Further selection experiments on industrial melanism in the Lepidoptera', *Heredity*, **10,** 287–301.
23 KLOMP, H. (1956). 'De terreinkeus van de Kievit, *Vanellus vanellus* (L.)', *Ardea*, **42,** 1–140.
24 KRUUK, H. (1964). 'Predators and anti-predator behaviour of the Black-headed gull (*Larus ridibundus* L.)', *Behaviour, Suppl.*, **11,** 1–130.
25 LANYON, W. E. and W. N. TAVOLGA (eds.) (1964). *Animal Sounds and Communication*, American Institute of Biological Sciences, Washington, D.C.
26 LILEY, N. R. (1966). 'Reproductive isolation in some species of sympatric fishes', *Behaviour, Suppl.*, **13,**.
27 LORENZ, K. (1931). 'Beiträge zur Ethologie sozialer Corviden', *J. Orn, Lpz.*, **79,** 67–120.
28 —— (1935). 'Der Kumpan in der Umwelt des Vogels', *J. Orn. Lpz.*, **83,** 137–213, 289–413.
29 —— (1962). 'Gestaltwahrnehmung als Quelle wissenschaftlicher Erkenntnis', *Z. exp. angew. Psychol.*, **6,** 118–65.
30 MANNING, A. (1956). 'Some aspects of the foraging behaviour of bumblebees', *Behaviour*, **9,** 164–201.
31 MARLER, P. (1957). 'Specific distinctiveness in the communication signals of birds', *Behaviour*, **11,** 13–39.
32 MAYR, E. (1963). *Animal Species and Evolution.* Harvard University Press, Cambridge (Mass.).
33 NICHOLLS, C. A., and D. A. ROOK (1962). 'Preparation of bees for consumption by a captive bee-eater' (*Merops ornatus*). *W. Aust. Natur.*, **8,** 84–6.
34 PATTERSON, I. J. (1965). 'Timing and spacing of broods in the Black-headed gull (*Larus ridibundus* L.)', *Ibis*, **107,** 433–60.
35 PERDECK, A. C. (1958). 'The isolating value of specific song patterns in two sibling species of grasshoppers (*Chorthippus brunneus* Thunb. and *C. biguttulus* L.)', *Behaviour*, **12,** 1–76.
36 PHILLIPS, G. C. (1962). 'Survival value of the white coloration of gulls and other sea birds' (unpublished Doctor's thesis, Oxford).
37 PITTENDRIGH, C. S. (1958). 'Adaptation, natural selection, and behaviour', in *Behavior and Evolution*, ed. A. Roe, and G. G. Simpson. Yale University Press, New Haven.
38 PRECHT, H., and G. FREITAG (1958). 'Ueber Ermüdung und Hemmung angeborener Verhaltensweisen bei Springspinnen (Salticidae)', *Behaviour*, **13,** 143–211.
39 ROTHSCHILD, M. (1962). 'Development of paddling and other movements in young Black-headed gulls', *Br. Birds*, **55,** 114–17.
40 RUITER, L. DE (1952). 'Some experiments on the camouflage of stick caterpillars', *Behaviour*, **4,** 222–32.
41 —— (1955). 'Countershading in caterpillars', *Arch. néerl. Zool.*, **11,** 1–57.
42 THAYER, G. H. (1909). *Concealing-coloration in the Animal Kingdom.* Macmillan, New York.
43 TINBERGEN, N. (1951). *The Study of Instinct.* Clarendon Press, Oxford.
44 —— (1953). *The Herring Gull's World.* Collins, London.
45 —— (1957). 'The functions of territory', *Bird Study*, **4,** 14–27.
46 —— (1962). 'Foot-paddling in gulls', *Br. Birds*, **55,** 117–20.
47 —— (1964). 'The evolution of signalling devices', in *Social Behavior and Organization among Vertebrates*, ed. W. Etkin. University of Chicago Press.
48 —— and W. KRUYT (1938). 'Ueber die Orientierung des Bienenwolfes

(*Philanthus triangulum* Fabr.) III', *Z. vergl. Physiol.*, **25,** 292–334. (Translated in Vol. I of *The Animal in its World*, p. 146.)

49 —— G. J. BROEKHUYSEN, F. FEEKES, J. C. W. HOUGHTON, H. KRUUK, and E. SZULC (1962). 'Egg-shell removal by the Black-headed gull *Larus ridibundus* L.: a behaviour component of camouflage', *Behaviour*, **19,** 74–118. (Reprinted in Vol. I of *The Animal in its World*, p. 250.)

50 —— H. KRUUK, M. PAILLETTE, and R. STAMM (1962). 'How do Black-headed gulls distinguish between eggs and egg shells?' *Br. Birds*, **55,** 120–9. (Reprinted in Vol. I of *The Animal and its World*, p. 304.)

51 WEIDMANN, U. (1959). 'The begging response of the Black-headed gull chick' (Mimeographed report of paper read at 6th Internat. Ethol. Conf., Cambridge).

52 WILSON, E. O. (1963). 'The social biology of ants', *Am. Rev. Entomol.*, **8,** 345–68.

53 —— (1963). 'Pheromones', *Scient. Amer.*, **208,** 100–16.

54 WYNNE-EDWARDS, V. C. (1962). *Animal Dispersion in Relation to Social Behaviour*. Oliver & Boyd, Edinburgh.

(From the Department of Zoology of Oxford University)

14
On Appeasement Signals (1959)

Soon after my pupil, M. Moynihan, and myself had begun a joint study of the behaviour of Black-headed Gulls in the year 1951, we were both greatly struck by the curiously rigid sequence of movements we were later to call pair formation displays or meeting ceremonies (Moynihan, **17**). On the approach of a female the male bird adopts the 'oblique' posture and utters a 'long call'. This leads to the female alighting next to him. Both birds then take up the 'forward' posture, usually placing themselves parallel, alongside each other—not, as in many male–male interactions, facing one another. Finally, first one, then the other, will jerk itself into the 'upright' posture and at the same time 'face away' from its partner (Fig. 154). This last movement (another name for it is 'head flagging') we found the most surprising. In our first description we called it 'turning the neck toward the partner'.

I vividly remember how one morning, having finished our early observation session, we walked home to our camp across the open saltings of Scolt Head Island, discussing the puzzling nature of this behaviour. What we were used to seeing in the courtship displays of most animals were always movements that showed off some highly conspicuous structure; numerous examples of this can be found in the works of such men as Selous (**19**), Huxley (**8**), Lorenz (**11**) and many others. However, the displaying of a white neck by a bird that was nearly all-white somehow did not seem to make much sense. Then Moynihan said, with some hesitation: 'Perhaps it is rather the other way round: perhaps the movement serves *to hide the face*'. All of a sudden I understood. I realised in a flash that Moynihan had said something tremendously important. But it needed a great deal of enthusiastic insistence on my part before my pupil was convinced of the far-reaching significance of his own idea. Never before had I realised why this kind of teamwork was often so fruitful: the collaboration of a younger man whose mind flows over with ideas and whose thoughts have not yet lost their flexibility, with an older

Fig. 154. Three stages of the 'meeting ceremony' of the Black-headed Gull. Above, male 'long-calling'. Below, male (left) and female in parallel bill-up 'forward' posture. Opposite, male (right) and female 'facing away'.

investigator who has the benefit of long experience but who may have become stuck in a groove (*'durch Fachkenntnisse getrübt'*). To me, Moynihan's suggestion was immediately convincing because through it, a great many other facts which had previously been vaguely uncomfortable suddenly fell into place. For the discovery of the appeasement component of courtship displays filled a gap of which I had somehow been aware, if only dimly, ever since Konrad Lorenz and Jan van Iersel had taken the first steps in unravelling the complex motivation of such displays. As is well known these contain, in many animals, a number of movements that closely resemble aggressive postures. And particularly since van Iersel had convinced me of the truth of one of Lorenz's earlier suggestions (which neither of us had at first believed), namely that the well-known zig-zag dance of the male Three-spined stickleback must originally have been (and up to a point still is) an alternation of a movement revealing a readiness to bite the female, and one showing a readiness to guide her to the nest, I had often asked myself why in that case the situation never, or at any rate rarely, ended in a real fight between the two members of the pair.

In recent years, we have been studying this phenomenon further, especially in gulls. It is the purpose of this paper to give a short outline of the complex problem, with special attention to its evolutionary aspects. How can such displays have originated, and what made them develop into their present forms? I am, needless to say, fully conscious of the fact that, while it may be possible to investigate certain aspects, e.g. causation and survival value of signals with the most rigorously critical experiments, the ultimate question of all research into evolution, i.e. the reconstruction of a past that is for ever hidden from our gaze, can at best receive a conjectural answer. The degree of credibility of such an answer will ultimately depend on the number of phenomena which fit into the suggested picture. In this short paper I shall try to present those arguments I consider the most important for my own conclusions and my way of looking at the problem. For the rest, I shall largely rely on the cooperation of my readers, and, following the example of Oskar Heinroth, I shall simply say: 'Here are the facts—draw your own conclusions.' ('*Den Reim dazu können Sie Sich selber machen*'.)

I start from the assumption which is now generally accepted in principle, that many 'expressions of emotion', whether or not they have a signalling function, result from activation of two, or even more, systems of movement at one and the same time. In order to place this first step in our analysis into the right perspective I want to remind you that many movements of animals are components of highly complex systems which owe their survival value, at least partly, to the fact that they form both functional and causal wholes. In a normal animal, the functionally related movements which constitute feeding, fighting, fleeing, grooming, mating, etc. occur together. Though we know very little about their causation, we do know that they are, in some way, causally connected with other movement belonging to the same system.

By analysing the motivation[1] (or total causation) of any display into components belonging to two or more such individual systems, we have admittedly taken only a first step towards its physiological elucidation—yet it is an essential and indispensable step.

It has long been realised that many 'mood-expressing' displays ('Ausdrucksbewegungen') are causally connected with several systems of movements. But that realisation cannot yet be called common knowledge because it still lacks scientific proof in the strict sense of the word. The reason for this is, partly, that many scientists, including zoologists, are slow to recognise the implications of published data; partly, that the phenomena themselves are not sufficiently well known; and finally, that we ourselves have not yet published our findings in a sufficiently objective, verifiable form. In fact I am con-

[1] In the context of my present attempt to analyse the complex cause of a display into systems that are one degree less complex I needed a word to denote the total readiness to carry out the display as measured by the occurrence of certain movements. I am using the word 'motivation' in a sense that is akin to that of readiness, mood, or tendency. It may be employed at various planes of integration; it is equally possible in this context to speak of a threat-motivation as of a (partial) withdrawal or (partial) attack-motivation.

vinced that we are not even quite agreed among ourselves which facts are to be used as bases for which conclusions, or why we are justified in drawing some of these conclusions. I therefore believe that an attempt to consider our methods of analysis, their potentials and their limitations, is called for.

It would seem that in analysing motivation we have so far employed three distinct methods.

1. Some expressive displays indicate their complex origin by their form. Elements of two or more behaviour patterns may be combined in three different ways: either (1) fragments of sub-systems may alternate, as in the zig-zag dance of the male stickleback (van Iersel, **9**), in the pendulum flight of many song-birds when they fight, or in the inciting display of some ducks (Lorenz, **13**); or (2) the components occur in a 'mosaic', simultaneously—as in the upright threat posture of gulls (Tinbergen, **22**; Moynihan, **17**) and in the facial expressions of canids (Lorenz, **14**); or else (3) the elements give rise to compromise movements (Andrew, **1**).[1] Obviously such an analysis is impossible without a thorough knowledge of the entire repertoire of possible movements of a given species. Much also depends on the ability of the observer to distinguish certain components or elements which might easily be overlooked as trivial; for instance I have only quite recently realised the importance of noting how far away from an adversary an animal will come to a halt, or how the longitudinal axis of its body is oriented when it does. Both these phenomena may be a measure of the degree of flight motivation present.

2. It is frequently possible to observe that certain displays regularly occur in alternation with quite different patterns of movement of which both the function and causation are largely known, whereas they occur rarely or not at all in association with other behaviour patterns. For instance all territorial and many other species possess a number of threat displays (i.e. displays with an intimidating or distance-increasing or -maintaining function); these alternate rapidly (often within seconds) with attack or flight, but not, as a rule, with other behaviours. The relevance of this fact lies in the experience that in a constant environment a dominant mood does not normally change suddenly into an entirely different mood. In an undisturbed animal, such moods build up slowly, and they also subside slowly ('inertia', after Lorenz, **12**). A difficulty, on the other hand, is the fact that in any natural conflict situation there is always the possi-

[1] It should be noted that Andrew's useful term is here given a new meaning; the parallel stance of courting Black-headed gulls has an orientation that is a compromise between 'facing' and 'turned away'.

bility that the threatened animal may influence the mood of the threatening animal in the direction of either flight or attack by the way it reacts to the threat. In other words, the only observations which are valid are those in which the external stimulation remains constant. There are two possible ways of overcoming this practical obstacle: an experienced observer, who knows from extensive previous observations which reactions of the adversary will influence the readiness to flee or to attack, will be able to single out sequences in which the adversary does not show any reactions likely to have these effects. In principle such a selective technique is no different from discarding a 'jump' of a barometer due to the slamming of a door. But we cannot be surprised if non-ethologists are not prepared to concede its validity. For these selective observations are admittedly difficult to make because they frequently involve split-second appraisals of the behaviour of two animals. Observers vary greatly in their ability to do this; and there is no doubt that wishful thinking might all too easily affect an observer's judgment. Hence, though it may be regrettable that so many scientists are unduly impressed by the exactness of their mechanical measuring instruments, and insufficiently appreciative of the potential performance of their own nervous 'measuring equipment', we must take account of this widespread attitude. It makes it not only desirable but essential for us to make our methods as objective and as verifiable as possible. In our studies of conflict behaviour we have therefore endeavoured to create constant conditions wherever possible, and one way of doing this is by replacing one of two interacting animals by a dummy, which we ourselves can control.

After Moynihan (**17**) had used the selective technique, another pupil of mine, G. Manley, studied certain signalling displays of the Black-headed Gull with the aid of dummies. Unfortunately, good dummies are difficult to produce, for Black-headed Gulls, like many other birds, will not accept dummies unless they are clothed in feathers. And as these experiments frequently end in violent attacks, our original dummies (stuffed birds) had soon to be replaced by new ones—a costly procedure. After the first series of dummy experiments, Manley said to me in a tone of mild disappointment, 'but these experiments do not tell me anything I had not already learned from field observations!' This remark was of course less a criticism of the dummy technique than evidence for the effectiveness of the technique of selective observation. Nevertheless it was not quite justified. The superiority of the dummy experiment lies in its being repeatable and controlled. It is possible to make statements that can be tested and compared, for instance: 'A stuffed gull in breeding plumage: beginning of the breeding period; placed for 10 minutes at a distance of 50 cm from the nest; releases in 10 presentations 12 pecking and 5 retreat reactions.'

3. Motivation of expressive displays may be judged by yet a third method. Once the situations most strongly conducive to the release of two given behaviour patterns are known, it is frequently possible to observe that a given display will occur in situations which are intermediate between the two optimum releasing situations. To give a concrete example: certain threat displays occur chiefly on the border between two adjoining territories. We know that the rival will be attacked on one side of the border, and given a wide berth on the other. This allows us to conclude that the border is intermediate between these extremes and that arrival in this place will simultaneously arouse both aggressive and flight tendencies. It must be stressed that here, too, it is essential for the observer to be fully conversant with the situations that bring about aggression or flight, but it should be equally emphasised that at this stage it is not necessary to analyse these situations further. For instance, for our present purpose it is quite sufficient to know that a male stickleback in mating attire stimulates another male to aggression on one side of the territorial border and flight on the other. The fact that the effective stimulus for both reactions is the red colour of the fish, and not its shape, is, for the moment, not relevant.

These three methods have often been used intuitively, but rather less often consciously and deliberately. I believe it is essential that they be employed in a great many individual observations under conditions of the utmost precision and repeatability; it was only in this way that we ourselves became aware of the differences between the conflicts which underly different displays.

Whenever we were in a position to apply all three methods, results were in good agreement—a fortunate, and highly encouraging as well as significant circumstance. *Formal* analysis of the upright threat posture of gulls revealed that it was the result of simultaneous stimulation of flight and attack; the observation of *sequences* of movements furnished the same conclusion; and the analysis of the *situations* in which it occurs confirmed it. This consistency of results of what were at the outset independent methods seems to me more significant than is often realised.

When one is engaged in the study of such highly complex systems of which, moreover, so few variables are known to us (let alone under our control) it is always advisable to apply all methods available. The three I have just mentioned are all concerned with the interpretation of phenomena presented to us by nature. But it is also possible to approach the problem by a more deductive method: we may try to find out whether the expected displays will also occur in conflict situations which we create artificially by presenting to the animal the causal factors governing the component movement systems. Various possibilities for such experiments exist. Provided the test animals are in the required condition it is possible, for instance, to stimulate both flight and attack at one and the same time by the use of suitable dummies. Many years ago I 'conquered' a male stickleback in his own territory with a dummy rival and was able to observe that, as soon as the flight tendency had subsided, the vertical threat posture and a full-scale attack followed in quick sequence (Tinbergen, **20**). At that time I was not sufficiently aware of the fundamental importance of this method of manipulating component systems. By a judicious alteration of internal and external factors, my pupil Dr Tugendhat was able to produce, in sticklebacks, conflicts between food-searching behaviour and flight. Her careful observations of the behaviour which results from varying levels of excitation in each system are in good agreement with my previous inferences (Tugendhat, **23**). We also expect interesting results from a judicious use of drugs that stimulate one system more than another, and there is no doubt that electrical stimulation of the brain will be useful for further testing and analysis.

Admittedly the progress achieved with all these methods has been painfully slow; yet some displays have now become much better understood. Threat postures for instance always originate from a conflict between aggression and flight. But it is important to give accurate descriptions of the activity systems involved. Thus it is far from irrelevant whether the sub-systems involved are termed 'attack' or 'approach'; 'flight' or 'moving away'. The more general term 'approach' would only be justified if it could be shown that *all* types of approach, irrespective of the motivation behind them, might be involved. For instance, the upright threat posture of gulls will never occur when a hungry gull is afraid of approaching its food because of the presence of an observer, or when a gull collecting nesting material dares not go near the observer's hide. Nor will it occur when a breeding gull clearly wants to leave the nest but is drawn to sitting or brooding by the sight of its eggs or its young. Grass-pulling displays (Vol. 1, Fig. 9) of large gull species will not take place in a conflict between fear and obtaining food; but they can be elicited by a sheep coming dangerously near the nest. Whenever they do occur, it is attacking and fleeing that alternate, and no other system is involved. These threat displays are thus not simply the outcome of a conflict between 'approach' and 'avoidance'; they are much more specific and occur only when there is a conflict between a tendency to attack and a tendency to flee. Many authors speak of 'approach-avoidance-conflicts' when, say, food gathering and fleeing are elicited. It is not always possible to infer from the observed facts whether such general terminology is justified. It is important to study each case very carefully in its own right. Kittiwakes turn their face away when they are stimulated to flee yet for some reason want to stay (Vol. 1, Fig. 12) (E. Cullen, **3**). Eider ducks lift their chins when they are frightened (stimulated to flight) but for some reason (e.g. either a sexual tie to the male partner, or a parental tie to the brood, or a tie to the feeding place) do not want to move from where they are. According to Andrew (**1**) the tail-flicking of some song birds is symptomatic of a conflict between a desire to fly and a desire to hop—both controlled by the same overall motivation. These conflicts are different from attack–escape conflicts.

The nature and intensity of a conflict affect the resulting movement in two ways. Again we know a little about this from threat displays. Even an inexperienced observer will notice that in a conflict between birds that are not quite evenly matched, the threat postures of the bird that is winning (but is not completely uninhibited in its attack) will differ from those of the bird which is losing but which is not sufficiently inferior to react with uninhibited flight. It is possible to determine the relative intensity of these tendencies more accurately by counting actual numbers of attacks and retreats in experiments with dummies. These show that there a number of ambivalent movements, which can be arranged in a series from the relatively more aggressive to the relatively more defensive threat displays. The latter indicate that the animal will fight only if it is attacked and cannot escape; the former indicate that attack will almost certainly follow. We also know that certain threat displays are associated with mild conflict and others with the very strong simultaneous activation of aggression and flight. Hence threat displays vary in two ways, by the *relative* and by the *absolute* level of flight and attack tendencies.

Obviously motivation of courtship displays, such as the initiation of mating by the male stickleback and the sequence of pair-formation ceremonies of the Black-headed Gull, is even more complex. In Black-headed Gulls (Moynihan, **17**) and certainly in many other territorial animals (cf. for instance Morris, **15**; Hinde, **6**, **7**) sexual attraction, aggression and fear are all involved. As Heinroth (**5**) had already described back in 1911, and Kortlandt (**10**) has stressed anew, a distinction must be made, at any rate in pair-forming species, between the sexual urge and the urge to pair formation. G. Manley, who studied the courtship display of Black-headed Gulls in detail, largely confirmed this distinction. But he also showed that the contribution of sexual urges is greater than Moynihan had assumed.

It does no harm to re-state the reasons which lead us to talk about aggressive elements of courtship displays rather than, the other way round, about sexual components in threat displays. Application of the three methods of motivation analysis has shown that the motivation of many courtship (male–female) display movements is really partly hostile, but that sexual components in hostile (male–male) behaviour are rare. Furthermore, we can observe that, if these components of courtship displays which we interpret as 'hostile' are very pronounced, as frequently happens, they are generally associated with a large amount of aggression or flight in such a manner that pair formation is impeded rather than promoted; and when the same display components are particularly pronounced in male–male confrontations they will enhance the effect of overtly hostile behaviour (i.e. increase the distance between two individuals) and could therefore hardly be associated with more successful pair formation. Psychoanalysts in particular are inclined to hold to the reverse interpretation. This may be connected with their general overemphasis on sexual motivation. I believe it to be a necessary though very difficult task to make them reconsider this idea on the basis of observed facts of the kind described here: psychoanalysis (and human behaviour studies in general) could profit greatly from giving more attention to non-verbal behaviour.

As I stated at the beginning of this paper, pair formation or meeting displays in Black-headed Gulls consist of the following sequence: 1, The oblique posture-with-long-call; 2, the forward posture; 3, the 'facing away' posture. I should now like to use an analysis of the 'facing away' display for a discussion of the concept of appeasement gestures.

This 'facing away' is not, as we first thought, confined to the pair formation ceremony. It occurs equally in male–male confrontations, usually in the animal that is less strong and fares less well in the fight. Though occasionally a bird that is actively attacking may run up to its adversary with its head turned away, we believe that even in those cases fear can always be shown to be a component in the total motivation. While in the male–male situations 'facing away' may be associated with any posture, it is in the pair-forming situation, invariably coupled with the 'upright' posture. But this is an aspect of ritualisation, a topic I must disregard for the time being. The fact that 'facing away' occurs in both the above situations is evidence for its being connected with flight on the one hand, and with approach—in hostile as well as sexual situations—on the other.

In Kittiwakes 'facing away' occurs whenever fear is coupled to any sort of desire to approach or stay. E. Cullen (**3**) observed it regularly in birds which, having usurped a nest, preferred to stay on the nest rather than to flee from its rightful owner. Kittiwake chicks attacked by their siblings or by other young gulls will face away if they do not want to move.

Sometimes it is the actual form of the movement which betrays its relationship to a tendency to flee. Weidmann (**24**), studying the Common gull, was able to observe a whole series of transitions, from turning away the head alone, through turning away the whole body, to, finally, flight.

In Black-headed Gulls, 'facing away' is more complicated; even though we must assume that it was originally an intention fleeing movement it can, in male–male encounters, be done by an attacking bird. I believe that in the series from highly aggressive to more defensive and timid displays, there are many transitional forms, and combinations which we do not yet understand. I shall come back to this phenomenon, which is of considerable importance for our understanding of evolution.

Unfortunately we have so far no experimental evidence as to the function of 'facing away'. For various reasons, our experiments with dummies in that posture were not very successful. However, long series of observations on kittiwakes and Black-headed Gulls indicate that the attacker is very suddenly and strongly inhibited by this signal. It is interesting that the effect of the 'facing away' movement (and probably of many other appeasement signals) is not identical with that of threat displays. It is true that threat displays frequently have the same effect of reducing attack; but they do so by stimulating flight (i.e. by intimidation) which in turn suppresses aggression. Facing away, on the other hand, does not induce flight in the aggressor; it appears to inhibit both flight and attack. It seems to do this by bringing about the disappearance of some of the stimuli that release aggression and withdrawal. According to Cullen's study (**3**) of the Kittiwake, it is the beak that triggers off aggression in the adversary and gives it direction. And in fact a Kittiwake will often try to get hold of the turned-away beak of its adversary by reaching sideways round its head. For in the Kittiwake the aggression of the attacker does not subside completely, it continues to manifest itself in loud calls and in the attempt, just described, to attack from the side. Or it may result in fresh attacks on some other gull, for instance on the female if she should happen to be present. It is characteristic of the effect of threat displays, and also of appeasement gestures that the aggressive mood remains but is no longer vented on the original adversary.

That is to say, the change is initially one of orientation and is followed by one of motivation.

One thing is certain: the facing away of the Kittiwake cannot be regarded as the 'offering of a vulnerable part' (Lorenz, **14**). Rather it should be described as the *concealing* of a structure that releases hostility. In threat displays the weapon, e.g. a bird's beak, is put into a fighting position and shown to the enemy—frequently in conjunction with bright colour releasers. Appeasement signals are, I believe, almost the exact opposite. I therefore believe it would be more exact to speak, with a slight shift of emphasis, of a 'non-display' of organs designed for attack or intimidation. Nor should the increased security gained by such highly vulnerable organs as the eyes be forgotten.

Before I proceed to a discussion of origin and ritualisation I should like to offer a few more observations for comparison with other species, though I am aware that the basis for the interpretation of such comparisons is still weak. Yet I feel I have to discuss them, for they could, and should, stimulate further research. In our study of gulls we have so far found four different postures—all occurring towards the end of pair formation displays—which we believe have to be interpreted as appeasement signals. (1) The 'facing away' posture is found in Black-headed Gulls and closely related species (e.g. Hartlaub's gull, the American Laughing gull, Franklin's gull, and Bonaparte's gull (Moynihan, **18**)) and the large gulls (though I did not recognise it in the latter at first). (2) Kittiwakes show the 'choking upwards' display with closed beak pointing upwards, and also the ordinary 'choking' display (Vol. 1, Fig. 7e) which is always performed with the beak open, displaying its bright orange mucous membrane lining, and pointing downwards, whether it occurs as a threat or as an early part in the pair formation sequence. (3) Little gulls tilt their heads sideways. In terns, which behave similarly (J. M. Cullen, **4**), this has the effect of at least partially hiding the black cap (presumably part of the threat display from the adversary—admittedly this explanation is not quite as satisfactory for Little gulls in view of their head coloration). (4) Finally we are informed by Bateson and Plowright (**2**) that the Ivory gull uses 'head tossing' as an appeasement signal both during fighting and as a greeting. A detailed analysis of aggressive and pair-formation behaviour of more species will undoubtedly show many interactions of aggressive and sexual components in courtship displays, and may well reveal the convergent occurrence of these four types of appeasement signals and may perhaps reveal others.

I shall now discuss the ritualisation of appeasement gestures before turning to their origin. By so 'peeling off' secondary adaptations it will be easier to elucidate the problem of origin.

If we look simply at the form of movements we can see that the degree of ritualisation varies greatly even within a relatively homogeneous group such as the gulls. All species so far investigated practise 'facing away', but, as Manley points out, it is only in the pair-formation sequence of species belonging to the Black-headed Gull group that this movement is performed (a) very frequently, (b) with a jerk, and (c) invariably in the upright posture. Presumably this is somehow connected with the acquisition of a dark face, which reinforces the effect of the threat postures. Admittedly *Larus novaehollandiae* and its close relatives do not possess dark faces. But there is some evidence that they had them once and lost them again when returning to a coastal habitat. The 'choking' display on the other hand is much more striking in Kittiwakes than in Black-headed Gulls, partly no doubt because it is done rhythmically and with an open beak.

In addition we have to take into account the possibility that the motivation may frequently have changed. I should like to discuss this in relation to the behaviour known as 'head tossing'. With the large gull species 'head tossing' (Vol. 1, Fig. 11) is clearly effective as a 'friendly' gesture. It accompanies the begging behaviour of females, and both sexes use it to initiate coition. With other species of gulls the same behaviour acts as a 'defensive threat display' or, in the course of pair formation, as an appeasement gesture. Now, as Manley shows in more detail, all species of gull express the urge to flee in a conflict situation by a lifting of the beak (cf. Tinbergen, **22**). So it is conceivable that 'head tossing' is in fact a ritualised lifting of the beak in which 'suddenness' and 'rhythmic repetition' have been especially emphasised. What originally was a 'defensive threat display' characteristic of all species of gull came, in many species, to signify mere 'friendly appeasement'. Similar changes in the connection between the underlying (mixed) motivation and the overt movement or posture may well have occurred in the adaptive radiation of many displays.

As already mentioned, some analyses of postures due to mixed motivation have indicated that threat displays may be motivated to varying extents by the desire to flee and the desire to attack. The same applies to appeasement signals as regards the relative contribution of fear and a readiness to stay. It cannot be doubted that during the course of evolution a (never more than defensive) threat display became emancipated into a gesture of appeasement. During such a process a 'typical intensity' for the display (Morris, **16**) would not only have evolved, but would also have changed as the motivation itself changed. Such developments would have occurred particularly in those species which use agonistic behaviour as a spacing-out mechanism. In these species 'hostility' might make pair formation more difficult, but it is precisely this hostility which would most easily lead to the evolution of signals denoting inferiority in a situation where two prospective sexual partners meet. The fact that these signals have an attenuating effect on both attack and flight must have aided their selection both in their new context (sexual) and in ritualisation as genuine unambiguous appeasement signals.

It is not surprising that there is so little reliable evidence on the evolutionary process of emancipation, on changes in motivation. After all, our knowledge of the motivation of ambivalent movements is slender and tentative; the results of the first few qualitative analyses made of a handful of postures in a handful of species are so meagre that we are hardly yet in a position to make comparisons.

And yet I feel that the overall picture that begins to emerge is fairly satisfactory and consistent. Of course it needs strengthening in many places but it does contain certain concrete hypotheses that can be tested.

We start from the fact that in many of the higher animals attack and flight constitutes highly developed and vitally important mechanisms of dispersion. Conflicts between the urges to attack and to flee, which are unavoidable especially in territorial species, lead to postures which function as signals and have been appropriately ritualised. Potential sexual partners will, on meeting, release not only sexual reactions but also, at the same time, fleeing and attacking. Their meeting will thus inevitably result in a three-cornered 'conflict of motivations' and in such a situation the weaker partner—usually the female—is likely to develop defensive threat displays and appeasement signals. These gestures are then subjected to selective pressures arising from the pair-formation situations, with the result that ritualised pairing signals and threat displays diverge, each evolving the characteristic motivation and form that fit the requirements. Since it is probable that most of the conspicuous signal structures had their origin in hostile contexts—as threat reinforcements—they had to be put out of action (i.e. hidden) when appeasement gestures were ritualised. In some few instances, coloured structures may have developed in the context of appeasement, for example the black band at the back of the neck in young Kittiwakes (E. Cullen, **3**). There are probably others, e.g. the neck coloration of the Pintail drake and the white plumes displayed in the bowing movements of night herons, which seem to contain an element of appeasement.

Comparisons between species, however, tell us even more as already mentioned, a great many appeasement displays are of a form which could be said to be 'the exact opposite of threat displays'. In several such appeasement or strongly defensive threat displays gulls turn their beaks away from their adversaries, i.e. not, as in aggressive threat displays, towards them. In principle, a bird can do this in only one of four possible ways: by pointing its beak upwards; by turning it away horizontally; by bending low; and finally, by simply withdrawing its head. In this last position the beak still points towards the adversary, but it is withdrawn as far as is possible without withdrawing the whole body. This latter appears to be the origin of the 'hunched posture' (Vol. 1, Fig. 10), while the first of these ways may have led to the 'choking upwards' movement and the second, to the 'facing away' display. Hence we are led inevitably to the following conclusion: because of the fact that there are only four possible basic methods of not pointing the weapon (the bill) at an opponent, appeasement signals resembling defensive threat displays developed independently (by convergence) in a number of rather different groups. Examples may be found in another publication of mine (Tinbergen, **22**).

It is tempting and would be fruitful to extend these comparative studies to those expressions of emotion that are cross-culturally constant in Man. It is highly probable that a study of smiling and similar expressive gestures in primates will show that these facial movements, which in ourselves and many apes and monkeys have reached a high degree of ritualisation (and some of which are usually friendly both in intent and in effect) can be traced back to defensive threat displays. It will be equally rewarding to analyse the whole complex system of meeting behaviour in a similar manner. One reason why it is so important to include Man in these comparative studies is that only in this way may we hope to discover parallels between what can be observed objectively and the accompanying subjective experiences. It would be particularly interesting to find the subjective correlates to the emancipation of appeasement gestures from threat gestures. 'Feeling hostile' and 'feeling friendly' towards someone are after all very different experiences, and yet animal studies suggest that they may be linked in a graded spectrum of subjective states, and certainly what we have so far learned from animals points unequivocally to the existence of 'hostile traces' in many 'friendly' utterances.

I have not been able to do more than skim the surface of this whole complex of problems, and some of my readers may well be disappointed by the weakness of my arguments. But the main purpose of my paper was to show that *methods* are available, and ought to be applied more systematically and critically than is often done. We have enough facts to justify the expectation that we shall be able to deepen considerably our understanding of ambivalent motivation, which occurs so often in the lives of animals and Man. But to achieve this will require close integration of studies of causation, of function, and of studies using the fundamentally different method of comparative research.

REFERENCES

1 ANDREW, R. J. (1956). 'Some remarks on behaviour in conflict situations, with special reference to *Emberiza* sp.', *Br. J. Anim. Behav.*, **4,** 41–5.

2 BATESON, P. P. G. and R. C. PLOWRIGHT (1954). 'Some aspects of the reproductive behaviour of the Ivory gull', *Ardea*, **47,** 157–72.

3 CULLEN, E. (1957). 'Adaptations in the kittiwake to cliff-nesting,' *Ibis*, **99,** 275–302.

4 CULLEN, J. M. (1956). Unpublished D. Phil. thesis, Oxford University.

5 HEINROTH, O. (1911). 'Beiträge zur Biologie, namentlich Ethologie und Psychologie der Anatiden'. *Verh.* **5.** *Intern. Ornithol. Kongr. Berlin* **1910,** 589–702.

6 HINDE, R. A. (1953). 'The conflict between drives in the courtship and copulation of the chaffinch', *Behaviour*, **5,** 1–31.

7 —— (1954). 'The courtship and copulation of the greenfinch', *Behaviour*, **7,** 207–32.

8 HUXLEY, J. S. (1914). 'The courtship habits of the Great Crested Grebe (*Podiceps cristatus*); with an addition to the theory of sexual selection', *Proc. zool. Soc. Lond.*, 491–562.

9 IERSEL, J. J. A. VAN (1959). *Ambivalent Gedrag*. Leiden (inaugural address).

10 KORTLANDT, A. (1959). 'Analysis of the pair-forming behaviour in the cormorant, *Phalacrocorax carbo sinensis* (Shaw and Nodd.)', *Proc. 15th Intern. Congr. of Zoology*, London, 1958, 839–41.

11 LORENZ, K. (1935). 'Der Kumpan in der Umwelt des Vogels', *J. Orn. Lpz.*, **83,** 137–213, 289–413.

12 —— (1939). 'Vergleichende Verhaltensforschung', *Verh. dt. zool. Ges.*, 69–102.

13 —— (1941). 'Vergleichende Bewegungsstudien an Anatiden', *J. Orn. Lpz.* (*Festschrift Heinroth*), **89,** 194–294.

14 —— (1953). 'Psychologie und Stammesgeschichte', in *Die Evolution der Organismen*, G. Heberer. Stuttgart. pp. 131–72.

15 MORRIS, D. (1956). 'The function and causation of courtship ceremonies', in *L'Instinct*, ed. P. Grassé. Paris. pp. 261–87.

16 —— (1957). 'Typical intensity and its relationship to the problem of ritualisation', *Behaviour*, **11,** 1–12.

17 MOYNIHAN, M. (1955). 'Some aspects of reproductive behaviour in the Black-headed gull and related species', *Behaviour*, *Suppl.* **4,** 1–201.

18 —— (1958) 'Notes on the behaviour of some North American gulls. III'. *Behaviour*, **13,** 113–31.

19 SELOUS, E. (1909). *Bird Watching*. London.

20 TINBERGEN, N. (1940). 'Die Ubersprungbewegung', *Z. Tierpsychol.*, **4,** 1–40.

21 —— (1953). *The Herring Gull's World.* Collins, London.

22 —— (1959). 'Comparative studies of the behaviour of gulls (Laridae); a progress report'. *Behaviour*, **15,** 1–70. (Reprinted in Vol. 1 of *The Animal in its World*, p. 23.)

23 TUGENDHAT, B. (1959). 'Studies of the effect of thwarting in sticklebacks', D. Phil. thesis, Oxford. (See also TUGENDHAT, B. (1960).) 'The disturbed feeding behaviour of the Three-spined stickleback: I. Electric shock is administered in the food area', *Behaviour*, **16,** 159–87.

24 WEIDMANN, U. (1955). 'Some reproductive activities of the Common gull, *Larus canus* L.' *Ardea*, **43,** 85–132.

(*From the Department of Zoology of Oxford University*)

15
Ethology (1969)

I

In the last half century we have witnessed the development, or rather the revival, of a science of animal behaviour that is now widely known as ethology. Like so many other scientists, ethologists find it difficult to characterise their science. They agree that (as expressed in the word ethology) the starting point of their studies is an observable —'habits', in the sense of movements—and it is relevant to an understanding of their approach that they are biologists. I myself like to define ethology as the biological study of behaviour, a formulation that mentions both the observable phenomenon and the method of study.

One of the forerunners, though because of lack of immediate followers not a founder of ethology, was Darwin (**9**, **10**). With remarkable foresight he realised that if his theory were to explain evolution of animal species by means of natural selection he had to apply it to all properties of animals, whether 'structural' or 'functional', and therefore could not ignore behaviour. His work contains much concrete material that we would now call ethological. While it was at that time impossible to know much about the 'machinery' of behaviour, Darwin's procedure could be characterised by saying that he treated behaviour patterns as *organs*—as components of an animal's equipment for survival.

The ethological aspects of Darwin's work were not followed up at once. Biologists after Darwin concentrated until well into the twentieth century on the elaboration and consolidation of a picture of phylogeny, i.e. of the course evolution must have taken, and for this they selected the most obvious observables: structural characteristics. Thus for a long time zoology at least concentrated largely on comparative anatomy.

Now and then isolated starts towards a biological science of behaviour were made by individuals, without, however, setting off an immediate, continuous development. Neither Morgan's work (**42**) on instinct and habit, nor Whitman's sophisticated comparative studies (**65**) of simple behaviour patterns, nor Jenning's studies (**28**) of the behaviour of unicellular organisms, to mention a few outstanding examples, really caught on. The explanation of this failure is undoubtedly complex: the existing branches of biology, to which genetics had in the meantime been added, attracted the available talent; there were relatively few biologists, and their science did not have much of a status even in scientific circles; behavioural phenomena were too complex to yield easily to exact description, let alone to experimental analysis; above all, perhaps, religious attitudes hampered scientific analysis of animal behaviour, so uncomfortably reminiscent of our own.

In the first quarter of the century, however, two biologists began to produce work that had a more immediate impact. In Britain Huxley (**25**, **26**) began to redevelop the study of bird behaviour, particularly mating behaviour, with a view to understanding its evolution by natural selection. He argued that success in reproduction depends on epigamic displays and structures combined, and that both were parts of signalling systems in which each sex partner in a pair bond exerted the selection pressures that favour increased effectiveness as signals. In Germany Heinroth (**18**, **19**), in a series of deceptively simple but in fact highly sophisticated publications, applied the methods of comparative anatomy to behaviour patterns, mainly of birds, stressing the fact that many behaviour patterns are typical of species, genera, and even families and larger taxonomic groups. He also demonstrated the fact that many of these behaviour patterns were largely 'innate', in the sense of being not or hardly modifiable by variations in the environment in which individuals grow up.

It was perhaps no accident that both men worked mainly on the behaviour of birds. Birds are on the one hand sufficiently highly developed to make their behaviour seem a special, challenging phenomenon (unlike the seemingly more machine-like movements of, for example, a starfish) but on the other hand are less likely than, for instance, monkeys to be identified with ourselves.

The impact of their work was at first confined to the more scientifically-inclined bird watchers. A steadily swelling stream of papers on bird behaviour began to appear, of which some were of a very high standard (e.g. Verwey, **62**). But the real start of a more comprehensive attack was made by one individual, Lorenz of Vienna, whom Huxley (**27**) was later to call 'the father of modern ethology'.

Lorenz, who already at that stage had an unrivalled first-hand knowledge of animal behaviour, was the first to try to assess what existing science could contribute to our understanding of the normal, natural behaviour of animals under undisturbed conditions. After the publication of two penetrating studies of the social behaviour of Jackdaws (**36**, **37**), and a more general study of social behaviour (**38**)—a paper that unfortunately reached the English-speaking reader in a much too condensed, truncated form), he published in 1937 (**39**) a critical assessment of the views of Spencer, Morgan, McDougall, and Ziegler, in which he showed that none of these could be reconciled with the new facts he presented. He then outlined what amounts to a modern version of Darwin's approach. In it he stressed the particulate nature of many behaviour patterns, and emphasised especially those relatively simple components that often appear at the end of a variable sequence of movements. During these introductory movements the animal seems to aim at finding the conditions in which these end acts, or consummatory acts (which he called 'instinctive acts') could be performed. In his largely inductive procedure he attached much weight to the following phenomena: the occasional 'misfiring' of instinctive acts under grossly abnormal circumstances; the phenomenon of 'explosive' or 'vacuum' behaviour shown when an animal has for a long time been denied the opportunity to perform an instinctive act; the evolutionary conservatism of instinctive acts, which makes them to reliable species and group characteristics comparable to structural characteristics; the internally controlled maturation of many behaviour patterns in the course of the development of the individual ('innate' behaviour); a learning process that he called 'imprinting' and compared with the process of inductive determination in the development of organs; and finally the mosaic-like integration of 'innate' and learnt behaviour components found in the adult animal as the result of unequal modifiability of these different components. Much of this was still highly tentative, and later research has led to certain modifications, elaborations, and corrections; yet this work was like a breath of fresh air and opened up vistas of unexplored territory. It is often forgotten (due in part to a certain assertiveness of Lorenz's style) that he explicitly wrote (p. 330):

'I do not see any danger in formulating my (perhaps rather heretic) views in this provocative, extreme way, as long as we are aware that they are working hypotheses that we should be prepared to modify if new facts force us to do so. One thing, however, I hope and believe to have shown convincingly: that the investigation of instinctive behaviour is not a subject for grand metaphysical ("*geisteswissenschaftliche*") speculation, but, at least for the time being, a task to be pursued by concrete experimental analysis.' (Free translation by me, N.T.)

It is interesting to see, in retrospect, how Lorenz's work gradually extended its influence, first into biology, then into psychology and psychiatry, and how, reciprocally, ethology responded to influences from outside. My sketch of this further development will naturally be a personal, and perhaps coloured, account.

Lorenz's new approach appealed at once to a small number of zoologists, namely those who, like myself, had already chosen to observe animal behaviour, and had been deeply disappointed when they looked for guidance at the psychological and the physiological literature. These observers too had been struck by the relative lack of modifiability of many behaviour patterns and their consequent conservative behaviour in phylogeny; by the apparent 'spontaneity' of much behaviour, which at that time seemed incompatible with reflex-theories; and by the fact that many behaviour patterns could often be seen to contribute decisively to survival, and to success in reproduction.

Lorenz's approach also appealed by its comprehensive nature—it was clearly a broad-fronted attack. Many more animal types were considered than before; ethologists aimed at describing and analysing all complex behaviour shown by animals in their natural surroundings; finally they were interested in function and evolution of behaviour as well as in its causation.

Before his work became more widely known, Lorenz (**40**) elaborated two aspects of his theory that had not been given special attention in his first comprehensive theoretical paper quoted above. Both are important, because they had a bearing on the physiology of behaviour and led later to various contacts with neurophysiological work. Many non-conditioned behaviour components have been shown to be responsible to 'sign stimuli'—relatively minor parts of the total input with which the animal's sense organs could provide it. The organisation of such behaviour patterns was obviously such that, under natural conditions, the sign stimulus was sufficiently characteristic of the natural objects to which the animal had to respond, to guarantee that the fitting response would occur. In some abnormal environments (viz. those that present the sign stimulus in a different context) this dependence on sign stimuli can lead to responses to the wrong object. This selective 'filtering' of sensory input, and therefore the effective sign stimuli, varies with the response shown, and often with the internal state of the animal. This type of evidence soon provided a link with a problem recognised as important by physiologists and cyberneticists: the centrifugal control (by 'gating') of the admission of input.

The second point of later contact concerned the internal, particularly central nervous control of behaviour. Independently, another individual 'heretic', the physiologist von Holst (**22**), had been demonstrating the existence of what he called spontaneous and automatic rhythms in central nervous activity underlying relatively complex behaviour patterns such as locomotion—movements not very much less complex than those that Lorenz had called instinctive acts. Von Holst's brilliant work in this field, which relied on the application of unconventional methods, culminated in his studies of 'relative co-ordination' of central nervous rhythms, and did much to arouse interest in the physiologically oriented aspect of Lorenz's work.

All the papers reporting on these new developments were written in German and published in German journals, and it is not astonishing that their influence remained at first confined to the German-reading parts of western Europe, particularly Germany and Holland.

II

The further development of ethology has been characterised by Koehler, who repeatedly pointed out that, like many new ideas, it went through three stages. At first, the new approach was largely ignored. Then, as its bearing upon wider issues began to be recognised, it met with vigorous criticism. This criticism led to further contact, and, through this, to a period of interchange of views and facts that marked the beginning of cooperation. This stage is often succeeded by one in which some notions are recognised as so self-evident that we tend to forget that they have not always been with us. In some fields of ethology we are entering this latter phase now.

The criticism was to be expected, because the notions developed by Lorenz and von Holst were in more than one respect contrary to prevailing thought, and so were bound to be unpopular. Von Holst's work was disturbing to those who were accustomed to think in terms of the reflex movement as the basic component of animal movement, as at that time most neurophysiologists were. Lorenz's notion of 'innate' behaviour clashed with the views of most American psychologists, who considered learning as the most important process controlling the development of behaviour of the individual—an over-emphasis that had undoubtedly much to do with the fact that animal psychology had derived its problems from human psychology.

Both Lorenz and von Holst expressed their views in an assertive, almost provocative way, reacting as they were against what they felt were strongly entrenched but biased positions. This made the ethological assault (and the counter-attack that it soon elicited) into an invigorating phase in scientific growth, characterised by a battle of wits fought with great intensity, from which, however, no side emerged loser or winner; all three disciplines profited by the confrontation.

The counter-attack opening the second phase came, after some skirmishes, when Lehrman published his 'Critique of the theories of Konrad Lorenz' (**33**), in which he made himself the spokesman of many of his colleagues among American psychologists.

Lehrman's criticism was in my opinion most incisive in questions of behaviour development. He argued that it is heuristically unhelpful to classify behaviour patterns, or parts of them, as either 'innate' or 'learnt'. Instead, the developmental processes ought to be analysed, even of those behaviours classified by ethologists as 'innate'. While such behaviour patterns admittedly developed without the aid of certain specified outside influences (for example when they were performed correctly before they could have been practised, or in response to stimuli which the animal had never met before), they might still be dependent on certain interactions with the environment, some of them different from those conventionally lumped under the name 'learning'. Lehrman claimed that the urge to disentangle such developmental processes is, so to speak, lulled to sleep by the rigid dichotomous classification of two behaviour types. Although he almost spoiled a good case by over-emphasising the part played by early experience and by under-representing ethological work on behaviour ontogeny, later research has justified a great deal of what he said.

With regard to the physiological analysis of behaviour, Lehrman argued that ethologists often confused similarity in achievement of unknown mechanisms with similarity in organisation of their machinery. While I think that he underrated the sophistication of the biologist precisely on this issue, his uneasiness was not entirely unfounded.

A third point of attack, related to both other points, was aimed at the optimism with which ethologists tended to assume similarity of mechanisms, both ontogenetic and physiological, between animal and human behaviour. As I hope to show later, modern ethologists, while still believing that Man and animals are more similar in their organisation than many psychologists are prepared to admit, acknowledge perhaps even more readily than many psychologists the extent of our ignorance with respect to human behaviour, and in addition have shifted the emphasis of their argument to a point of methodology: ethologists now claim that they have developed *methods* which could fruitfully, and in fact should, be applied to human behaviour.

Finally, Lehrman showed the weakness of the evidence on which ethologists based their conclusion that autonomous processes in the central nervous system play a prominent part in the causation of behaviour. He made clear that they ignored the possible influence of one or more peripheral, in part sensory influences (for example proprioceptors, i.e. sense organs that report on the conditions and events inside the body rather than on the environment), which studies of intact animals could not properly elucidate.

Contact with physiologists developed a little later. The main reason for this seems to me to be the fact that both subject-matter and methods of the two sciences were originally too different. Neurophysiologists were at that time very much occupied with the analysis of processes at the level of the neurone or relatively simple neurone systems, and neither Sherrington's *Integrative Action of the Nervous System* (**50**) nor Pavlov's work could be directly applied to complex behaviour. Von Holst, himself a pupil of the physiologist, Bethe, appealed at that time more to ethologists than to physiologists. Closely related to this difference in subject-matter was the difference in method. Physiologists were experimenters, and as such relied almost exclusively on the study of consequences of interference; as von Holst used to remark, this contains a methodological bias because by its nature the method deflected attention from what animals do when not interfered with, and thus from 'spontaneous' behaviour. Ethologists, faced with a paucity of purely descriptive data, tended to devote a large proportion of their time to observation, without as yet following up the resulting hypotheses with cogent experimentation. The gap, due to these differences in subject-matter, in method, and consequently in concepts and terminology, was wider than even those realised who saw the basic similarities in approach and believed in the ultimate fusion of all types of 'physiology of movement'. The early 'physiologising' by ethologists was undoubtedly naïve; for instance the relevance of 'rest output' of sense organs to problems of 'spontaneity', and that of proprioceptive input to problems of drive reduction was not grasped, nor were we in general aware of the important concept of feedback. Reciprocally, it did not help our understanding of behaviour mechanisms when, for instance, Pavlov (**43**) coined concepts such as 'the curiosity reflex' or when later Eccles (**13**), on the basis of neurone studies, elaborated on the 'mind'.

It was not astonishing therefore that contact between physiologists and ethologists developed relatively late, and that, when neurophysiologists began to take notice, they did so in a very critical state of mind.

The sharp distinction made here between psychology and physiology is of course highly schematic, particularly since American psychology had, by the time it came in contact with ethology, already developed physiological approaches and techniques. Yet the distinction is roughly valid because the centre of interest of psychologists was behaviour and that of neurophysiologists the nervous system.

Thus lack of experimental sophistication of ethologists, their concern with phenomena that were outside the sphere of interest of either psychology or physiology, and their interest in problems of survival value and evolution made it almost inevitable that their work elicited criticism from their fellow scientists; but it was equally inevitable that much of the criticism was based on misunderstanding. Under these circumstances, which could easily have led to a hardening of attitudes and mutual rejection of views and facts, it has been encouraging to see how soon the phase of *voicing* criticism moved into one of *listening* to criticism and, through this, to both a broadening of approach in all sciences concerned, and to convergence and fusion.

III

The post-war period has seen a rapid development of all the sciences concerned with animal behaviour. This development was undoubtedly in part due to trends in each of these sciences themselves. Psychologists were independently extending their interest to more types of animals, to more problems, and to the development of new techniques. Neurophysiologists began to extend their studies to higher levels of integration. Ethologists developed an increased tendency to experiment. It is worth stressing that this too was to a large extent due to Lorenz's work. As one who has been closely associated with Lorenz since the mid-thirties, and who is experimentally inclined, I can testify that, although Lorenz is himself not an experimenter, his way of formulating his hypotheses has inspired and even dominated much experimental work, and is doing so even now, though often so indirectly that the younger generation does not always realize this. I think I can characterise his influence best by saying that, while the experimental method would have been applied to behaviour problems even if he had not provided his hypotheses, the approach would have been far less system-oriented and more atomistic in character. There is also no doubt at all that through his penetrating comparative studies he gave great impetus to research into the evolution of behaviour, including the effects of natural selection—a field still practically ignored by neurophysiologists, comparative psychologists, and human psychologists.

But a striking aspect of the post-war development has been the start of interdisciplinary contact: ethologists, while still continuing to call attention to what they are convinced is new in their work, are absorbing a great deal from sister sciences. Perhaps the most obvious development is the growing interest in studies of causation, which naturally moved in the direction of behaviour physiology. But ethology has retained a breadth of interest that is less pronounced in the other disciplines, and even in other biological fields. I believe that non-biologists, and even a number of my fellow biologists, do not always realise the full width of the major problems that biology, in all its ramifications, covers. With a slight modification of Huxley's reference to 'the three major problems in biology', viz. causation, function, and evolution, I have always found it helpful to think of biology as concerned, in 'commonsensical' terms, with two problems: that of causation and that of function in the sense of survival value. By this I mean that, starting from observable life processes, we ask 'what makes this happen?' and 'how do the effects of what happens influence survival (including reproduction)?' The first question can be roughly divided into three separate questions, differing in the time scale involved. The causation of short-term, cyclical events in an animal's life (such as periodic feeding, or the alternation between reproductive and non-reproductive seasons) is studied with a view to analysing the causation, the 'machinery' of such cyclical behaviour. Over a longer stretch of time, the whole life of an individual is considered to be one long cycle, and the changes in behaviour machinery in the course of the cycle, the development of the individual, has to be studied. Finally, as the generations follow each other, evolution takes place, in the course of which the ontogeny changes in successive generations. In spite of a certain degree of overlap, causation studies thus refer to three rather different problems. The study of the two former problems differs from the third one in that both are concerned with repeatable phenomena, which therefore can be submitted to experimentation with controls. But evolution is one single cycle, in which each step has been unique, and its study is therefore ultimately dependent on indirect methods, on reconstruction of what has happened. This is true even though, acting on the 'principle of actuality' (or constancy of the basic mechanisms throughout the past), one can study single components of the evolutionary process experimentally.

The problem of the functions served by life processes is in a sense of a different order. Organisms, 'building up negative entropy', seem to defy physical laws; they maintain themselves in what seems at first glance an unstable state. The fact that animals survive, reproduce, and evolve in the face of many opposing pressures makes us ask how they manage to do so. If survival of these systems—organisms or communities—were the observable from which we start, biology would be concerned with causation only, but since our observables are life processes, we have to follow up the cause–effect relationships in both directions; looking back in time when studying causation, forward when we study effects, in order to understand how these effects contribute to survival. Because, since Darwin, the organisation of animals is assumed to be the result of natural selection, which favours the more successful forms, the investigation of the way in which life processes contribute to survival forms part of evolutionary study as well, and biology is therefore concerned with effects of life processes for two reasons. Of course, whether one traces causes or follows effects, the method employed is the same: the unravelling of cause–effect relationships, ultimately by experiment.

In a science such as ethology, which is characterised by an observable phenomenon rather than by, say, an animal type (e.g. ornithology), a problem (physiology), or a method (experimental embryology), an interest in all these problems is natural, and in this broad biological interest ethology still compares favourably with the bulk of physiological and psychological work.

In studies of the *causation* of short-term cycles (behaviour physiology) Lorenz's original views have been partly elaborated, partly corrected, partly supplemented by notions that allowed new questions to be asked. Lorenz described how functionally unitary behaviour sequences (such as the different activities that together are called feeding) often consist of an introductory phase, which (following Craig, **6**) he called 'appetitive behaviour' and is characterised by a high degree of 'spontaneity' (i.e. internal initiation), by irregularity or variability of motor patterns employed, and by the fact that it is continued until the animal meets with a situation that elicits the next link in the behaviour chain. Thus a hungry animal sets out 'in search of' food, and the switch to hunting, eating, etc. occurs when food is actually perceived. He further called attention to the fact (likewise pointed out by Craig) that such sequences often end in relatively stereotyped end acts, or consummatory acts ('instinctive' acts), and that the readiness to repeat or continue the same type of behaviour drops more or less suddenly with the actual performance of the end act. He thus stressed the facts that behaviour is in part internally motivated; that activities that belong together functionally are also connected causally in such a way that they occur together in certain sequences; and that the entire complex, so to speak, loses its internal motivation after performance of the end act(s). This led him to introduce the concept 'action specific energy', a more or less metaphorical term for largely unknown agents that influence specific, functionally distinct behaviour systems rather than general activity. His studies of fluctuations of the readiness to perform a certain behaviour showed that if an animal is denied the opportunity to show such behaviour, it becomes progressively more ready to respond to incomplete stimulation, and when stimulated it shows increasingly complete behaviour. In extreme cases, behaviour that is normally dependent on elicitation by stimuli, will appear 'spontaneously' ('vacuum activities'). Lorenz went one step further and identified 'action specific energy' with one particular type of internal control, viz. control by processes in the central nervous system itself, which he thought might be akin to the 'automatisms' in von Holst's sense. Rather ignoring the possibility that the drive-reducing effect of performance of the end act might occur through the medium of, for instance, proprioceptive stimuli, or even outside stimuli, he ascribed drive-reduction to direct depletion of 'action specific energy'. The model he proposed as an analogy did much to establish an image of his theory that was narrower and more rigid than the theory itself. It consisted of a tank slowly being filled with water ('action specific energy'), which could be emptied through various outlets, each of which could be opened by a trigger device ('releasing stimuli') that allowed the water to run out ('consummatory act'). If no triggers were pulled, the water would overcome the resistance blocking an outlet ('vacuum activity').

Lorenz's emphasis on internal behaviour-specific determinants has undoubtedly been of great importance, even though his specific hypothesis was too narrow. For instance, new work on the 'rest output' of various sense organs—an output affecting the central nervous system even when no specific stimuli are administered—indicates that even specific behaviour patterns such as coordinated swimming may depend on such unspecific input. The exact source of input need not be important, but when it is completely absent swimming stops (e.g. Lissmann, **35**). Closely related to the notion of spontaneous central nervous rhythms is the idea that the well-known influence of certain hormones on behaviour is a direct one. Later analyses have shown that this is not invariably true: reproductive hormones may cause growth processes, for instance of peripheral sense organs, which as a consequence become capable of stimulating the nervous system. However, direct effects of hormones on the central nervous system have been demonstrated, and the main area of impact has even been identified as the posterior hypothalamus (Harris, Michael, and Scott, **16**). At the same time, neurophysiologists have demonstrated in an increasing number of cases that central nervous tissue does much more than meekly wait to be prodded into action. The old notion is being replaced by one envisaging continuously active populations of nerve cells that are usually inhibited, but when disinhibited by other centres or by sensory stimuli give rise to coordinated behaviour (see, for example, Roeder, **45**). This level of activity depends on a variety of (partly extraneural) agents, and the 'endogenous' contributions of the central nervous system are now being studied in terms of the relationships, often the discrepancies between the inputs and the outputs—in short on the *processing* done by the central nervous system.

Similarly, Lorenz's emphasis on the drive-reducing effects of performance has stimulated research, and through this has led to a more comprehensive view of the problem than he presented at the time. In particular, his 'psychohydraulics' model differed in one important respect from the biological systems on which it was based: its self-regulatory properties were not sufficiently emphasised. Under the influence of cybernetics (which reached ethology at second hand, via engineering science and physiology) it was recognised that, when performance of the end act 'as such' reduced drive, performance must provide some kind of negative feedback. Once the search for feedbacks began to be applied to major behaviour systems, it was found that feedback loops, involved in drive-reduction after performance of a consumatory act, could vary widely from one case to another. Thus satiation after feeding is at least in part under the control of negative feedback from one part or another of the intestinal tract (e.g. Berkun, Kessen and Miller, **2**; Dethier and Bodenstein, **12**); the reduction of sexual behaviour after sperm ejaculation was shown in a fish, the stickleback, to be a consequence of stimuli from the eggs rather than from the loss of sperm (Sevenster-Bol, **49**), a particularly extensive feedback loop since the eggs are laid by the female in response to behaviour of the male that precedes actual sperm ejaculation. In these cases, the animal responds to 're-afferent' stimuli, i.e. stimuli that report back on the effects of the activity performed. On the other hand, the autumn migration of starlings was shown by Perdeck (**44**) to end after the birds have flown a certain distance or a certain time, not when they have reached a certain end situation. Blest (**3**) analysed, in a moth, the reduction of the readiness to fly after having flown, and found, by elimination of all conceivable extra-neural feedbacks (including those from proprioceptors) that the feedback must arise within the central nervous system itself—an example even more fully in accordance with Lorenz's ideas than that studied by Perdeck.

The analysis of the organisation of complex, functionally unitary behaviour patterns is likewise being pursued and leads to a more complex and varied image than that originally proposed by Lorenz. The basic fact that functionally related behaviour components are appearing in certain orderly patterns has been established with much greater accuracy, one of the methods being applied being factor analysis (Wiepkema, **66**). In general terms this type of integration of functionally related components into systems means that the components must be causally related *somehow*, but the organisational principles involved are much more varied than that suggested by concepts of unitary drives in the sense of super-ordinated motivational factors controlling directly all components—perhaps with relatively simple threshold differences, or different relations with qualitatively different stimuli for each component. The web-like character of the mechanisms underlying major behaviour systems is now being shown to be more complex; they include many feedback relationships and mutual (facilitating and inhibiting) interrelations at and between a variety of levels. The work on the organisation of reproductive behaviour in birds (Lehrman, **34**; Hinde, **20**, **21**), for instance, has now moved far beyond the demonstration that reproductive hormones are required for their appearance. Hormones influence growth processes not only in other hormone-producing organs but also in sense organs and in effector organs, with numerous feedback links and interrelations between separate cause-effect chains. Conversely, the activity of the hormone-producing glands can be affected by external stimuli, part of which in turn arise from earlier hormone-controlled effects. Thus the endocrine cycle of female pigeons is affected by stimuli provided by a courting male. Nest building in canaries, controlled by the female sex hormone oestrogen, leads to the completion of the nest cup; this not only stops this phase of building but also reduces the further secretion of oestrogen. In the meantime the broodpatch on the female's underside develops under control of oestrogen together with a secondary hormone, and as a consequence becomes more sensitive to touch stimuli. This makes it send back strong touch stimuli when the female sits in the cup, and this again makes the bird switch to softer material, i.e. feathers, with which the nest cup is then lined. The diagram by which Hinde (**21**, p. 302) summarises the relations so far known or suspected, in both a positive and negative sense, between the hormones, the behavioural and the physiological changes, and outside stimuli, shows a highly complicated web.

The influence of reproductive hormones can thus be exerted in indirect ways, for example through growth processes, but also directly, as hormone-implantations in the brain have been shown (Harris, Michael, and Scott, **16**). Brain-stem stimulation experiments further show that electrical stimulation of very simple rhythms can produce behaviour patterns of great complexity, suggesting hierarchically organised executive systems (e.g. von Holst and von St Paul, **24**).

It is difficult to assess what contribution the concept of 'action specific energy' has made to these developments. It has certainly drawn attention to problems of the internal control of major behaviour patterns, and although it was clearly an exaggeration and simplification to ascribe so much to more or less autonomous central nervous mechanisms, both the 'spontaneous' and the integrative activities of the central nervous system are clearly much more important than was recognised at the time. The concept of unitary drives has further rendered excellent services in the first steps of the analysis of 'conflict behaviour', i.e. behaviour such as 'threat' and 'courtship' resulting from the simultaneous elicitation of two or more major behavioural systems (e.g. Tinbergen, **57**). For purposes of more detailed analysis, however, knowledge of the organisation of functionally unitary major systems becomes increasingly important.

Increased attention is also being given to the problem of internal control of sensory input. It has long been recognised that the responsiveness of animals to external stimuli varies, and must be under internal control, but analysis of the processes involved has not started in earnest until recently. Physiologists now begin to trace the messages going out from the centres towards the sensory periphery which can 'set' the sensitivity of sense organs and can even 'filter' part of the sensory input. Such centrifugal control of sensory input by inhibition may well be effected in a variety of ways and at different levels; in the ear of the cat it seems to be the nervous cells in immediate contact with the sensory cells which, by the outgoing nerve known as 'Rasmussen's bundle', can be prevented from passing on auditory input (Desmedt and Monaco, **11**).

Equally interesting is the widening approach to 're-afference': the processing of stimuli from the sense organs which are received as a consequence of an activity of the animal, and by which it judges what effect the movement has had. Of particular interest are those studies in which the interaction between centrifugal setting and re-afference is investigated. The re-afferent stimuli have much wider effects than that of reducing 'drive'; they are, inside the animal, compared with something like a 'template', a representation of what these stimuli 'ought to be'. It is the *relation* between template and actual re-afference that dictates which behaviour shall be shown subsequently. Von Holst and Mittelstaedt (**23**) have demonstrated by a simple experiment that such a template can represent quite complex features, and also that it can vary with the internal state of the animal. This experiment is worth being described in some detail. When an animal moves in relation to the visual environment (either because it moves itself or when the surroundings move) the image of the environment moves over the retina of the eye. When the animal is stationary and receives such a stimulus, it responds by correction movements (the 'optomotor response'), which restore its original position in relation to the environment. However, when in a stationary environment the animal moves on its own accord, the retinal image moves as before, and yet the animal does not then respond by moving back. Von Holst and Mittelstaedt showed in a hover fly that in the latter case the optomotor response is not simply inhibited, but that in both cases the animal receives re-afferent messages from the eyes. Correction movements follow only when these 'checking-up' stimuli do not tally with what could be expected: when the animal is stationary, retinal movement is not expected and must mean movement of the environment, but in the moving animal the retinal image *should* move, and the centres take this into account. In other words, at each moment these receiving centres are informed about the '*Sollwert*', i.e. the value that the re-afferent stimulus 'ought to' have—in functional terms: its 'neutral value', the template into which this re-afference should fit; actual re-afference and *Sollwert* are *compared*, and discrepancies between the two lead to correction movements. The experiment shows that the value of such a template is 'set' to correspond with movements that the animal is performing. Similar processes can easily be shown to occur in our own visual system, and they are no doubt of very general occurrence.

These and many other investigations are exploring a field that lies, so to speak, half-way between ethology and neurophysiology. Other such researches concern the integration of separate sensory data into unit-messages of a higher order, such as is the case in vision of form (e.g. Sutherland, **52**; Waterman, Wiersma and Bush, **63**) and of movement (e.g. Hassenstein, **17**). The result of this widening of the front to attack is that the distinctness between some fields of ethological, psychological, and physiological research is beginning to fade. This was to be expected, because these phenomena of different levels of integration are being studied with essentially the same method—

although of course the special concepts, techniques, and terminologies remain different, adapted as they have to be to the level studied.

Studies of behaviour *ontogeny* have likewise made considerable progress, but there is still profound disagreement between American psychologists and (particularly German) ethologists on the methods of approach, and even on the aims of research. This is clearly demonstrated in two recent treatises by prominent representatives of these two groups, Lorenz (**41**) and Schneirla (**46**). English ethologists occupy on the whole a middle position.

The dispute started when Lorenz (**37, 38**) described many examples in which animals show species-specific, complex behaviour patterns (often in response to specific stimulus situations) in circumstances that had precluded the opportunity either for learning by practice or example, or for conditioning to the effective stimuli. He criticised the tendency among American psychologists to assume that learning processes, and other types of interaction with the environment during growth, were the main or even exclusive determinants of ontogeny. By classifying behaviour patterns in two ontogenetic types, innate and acquired behaviour, Lorenz stressed the importance of internally controlled development.

In subsequent work, the two parties applied different methods. On the whole, the ethologists proceeded by demonstrating that certain learning processes, at first glance likely to take part in the moulding of such responses, were actually not involved in a number of examples. Thus Grohmann (**15**) showed that the incipient, gradually improving flight movements made by nestling doves were not influential in the development of the basic flying ability of fledglings; likewise many responses to intra-specific signals were shown to be unconditioned (summary, for example, in Tinbergen, **56**); and many more similar examples are known. Psychologists naturally studied the effects of various manipulations of the environment in which animals grew up, and, equally naturally, tended to emphasise those cases in which their search was rewarded. Thus, undeniably, ethologists studied a broader spectrum of phenomena, whereas psychologists penetrated more deeply into problems of learning.

Closely related to this was a difference in aims, which has only gradually become clear. Already Lehrman (**33**) argued that the analysis of the whole developmental process, from very early stages on, and with reference not only to learning but to other formative outside influences as well, should be the real aim. Lorenz made it clear recently (**41**) that '*Not being experimental embryologists* but students of behaviour, we begin our query, not at the beginning of the growth, but at the beginning of the function of . . . innate mechanisms' (p. 43). (Italics mine, N.T.) With Lehrman and many English-speaking ethologists I believe that we *should* be concerned with the entire developmental process—that behaviour development will only be fully understood if we apply all the methods of experimental embryology to problems of behaviour development.

In view of these differences of attention, of aims, and of methods, no useful discussion is possible without a prior consideration of some matters of semantics.

Male Three-spined sticklebacks, raised in isolation from fellow members of the same species, show, when adult, normal fighting behaviour in selective response to other males (e.g. Cullen, **86**). This is, in Lorenzian usage, an 'innate response'. Strictly speaking, the term here means a response neither influenced by the example of an adult male (a possible 'teacher'), nor by practice, nor by conditioning. It means a *non-learnt* response, and this, as Beach (**1**) has said long ago is a negative definition. To workers interested in the entire developmental process, the word 'innate' implies no interaction with the environment at all; not 'acquired' through such interaction, at any stage, at any level. Now it has been shown for instance in tadpoles that exposure to light at an early stage is necessary for the rods in the retina (one of the two types of light sensitive cells) to develop their proper function (Knoll, **29**). While this cannot be called learning, it *is* an interaction with the environment, which *is* required for the proper development of this component of the visual system. In the stricter sense, therefore, none of the subsequent visual response of the tadpole, nor of the adult frog, can be said to be innate, even though many of these later responses may not require learning processes. As long as the earlier stages of development have not been investigated, one has not eliminated possible interactions with the environment. This kind of gap in the evidence seems to me to be at the root of the conviction of many psychologists and English-speaking ethologists that it is not helpful to use a rigid dichotomous classification of innate and acquired *behaviour*, unless one qualifies the word 'innate' rather drastically, as indeed Lorenz does.

In other contexts however the word 'innate' is not objected to. When two groups of animals grow up in the same environment (admittedly a requirement that is difficult to meet in practice) and develop different behaviour repertoires, these *differences* are in the strictest sense of the word innate; to be more precise, they are due to genetic differences between the groups, and the ultimate aim of the ontogenist is to analyse where in the developmental processes of both, and how, the different genetic instructions lead to different results. The same formulation can be applied to similarities between animals growing up in different environments.

The dichotomy is further (as Lorenz has pointed) completely justified with respect to the *sources of 'programming'* of the animal. The information comes either from within the animal or from the outside world. Male ducks of many species have to become conditioned to the females of their own species at an early age ('imprinting'), but females of these species mate preferentially with males of their own species even if they have been raised with males of another species (Schutz, **48**). Of course internal programming is in its turn the result of the trial-and-error interaction with the environment which directs evolution and which we call natural selection.

It is very remarkable indeed that in this confused situation the actual research done by ethologists and psychologists is so very similar. The information, referred to above, that tadpoles require light for their visual system to develop fully, comes from a zoologist; a great deal of work on the genetics of behaviour differences is done in American psychology laboratories. This must mean that much of the disagreement is due to differences in formulating aims and conclusions, and in differences of degree in the guiding interest, which expresses itself in the concentration of psychologists on environmental aspects and of ethologists on internal aspects.

The procedure now applied by many workers could be roughly characterised as follows. Descriptive studies of the behaviour of developing animals reveal how the behaviour 'machinery' changes in the course of time. By comparing the effect of different early environments on the development, certain environmental features are either demonstrated to be effective or shown to be uninfluential. The subsequent analysis of the processes involved takes different courses for environmental effects than for internal aspects of programming. I think it is often forgotten by champions of 'innate behaviour' that, in order to get away from demonstrations that certain environmental aspects are *not* influential, one has to analyse internal processes directly, and this can only be done by interfering with processes inside the animal. On the other hand, champions of 'acquired' behaviour often seem to forget that 'acquisition', including learning, is not, so to speak, creating something out of nothing; it is a process of *changing*, and often perfecting, through interaction with the environment, something less perfect that was already functioning before. Thus the social interaction between a mother cat and its kittens begins with certain unconditioned responses in both, which improve by a series of adaptive adjustments made as a consequence of mutual interaction (Schneirla, Rosenblatt, and Tobach, **47**).

Another example is the development of song in chaffinches and other birds. Males raised without being able to hear the song of older, experienced males do not develop the normal song, but they do develop a simpler, warbling song (Thorpe, **54**).

As has been already indicated in the example of the tadpoles, it is further of great importance to distinguish clearly between stages of development and between the levels of integration involved.

The relevance of such facts to our problem could perhaps best be made clear by the following analogy. If, in the course of constructing a piece of scientific apparatus, I ask a factory to cut a sheet of metal into the desired shape, which I subsequently bend before incorporating it into my apparatus, is this apparatus factory-made or home-made? With respect to the metal part itself, would it make sense to call *it* either factory-made or hand-made?

A few examples must suffice to show how a picture is gradually emerging that shows promise for our ultimate understanding of the developmental process; a picture much richer than a mere dichotomy into innate and acquired responses suggests. This is not to say that Lorenz's emphasis (**41**) on the extent of internal programming has not been important—in fact I think that, particularly with the extension he has recently added by introducing the concept of the 'innate teaching mechanism' (see below), it will have very far-reaching consequences even for our understanding of human behaviour. But it is worth pointing out that we have so far not agreed on a sensible set of concepts and terms.

The considerable promise of direct attacks on internal processes has been demonstrated by the rightly famous experiments of Sperry and his collaborators (see, for example, Sperry, **51**). In the normal development of a vertebrate the sensory nerves that report messages from the skin grow out from the spinal cord towards the skin areas they serve. When, at an early (yet not too early) stage of development of a tadpole a piece of prospective dorsal skin tissue is transplanted to the ventral side of the embryo, it develops into a patch of dark-coloured, i.e. dorsal skin, so that the adult frog has on its belly a piece of skin that has the characteristics of a piece of dorsal skin. Conversely, prospective ventral skin that is transplanted to the back (and in practice both manipulations are done at the same time by exchanging the two pieces of tissue) likewise retains its characteristics, and so appears later as a piece of light-coloured skin on the back. When these transplantations are carried out before the sensory nerves have grown out to the skin, the nerves that establish contact with the grafts are those that would normally have contacted the skin areas now occupied by the transplants. When now the adult frog is touched on one of the transplanted skin areas, it makes, as any normal frog does, scratching or wiping movements with a hind leg. But unlike normal frogs, such a frog wipes its *ventral* side when the piece of white skin on its *back* is touched, and it wipes its *back* when the piece of dark skin on its *belly* is touched. The conclusion is inescapable that the nature of the skin, even when it grew up in the wrong place, determined the later function of the sensory nerve and the reflex mechanism it serves. The stage in the developmental process in which the function of the nerve is determined is therefore under internal control with respect to the animal as a whole (although, since the skin exerts an influence on the nerve, we cannot call the process 'internal' with respect to the central nervous system). Sperry suggests that similar internal formative processes may well take place in a staggered series of events inside the central nervous system itself.

Since the development of young animals cannot be sharply distinguished from such processes as the seasonal development of the reproductive condition, the work of Hinde and Lehrman mentioned before, which is concerned with another set of internal determinants, the hormones, is relevant here too. Hinde's work in particular shows how fruitful it is for such studies to analyse the interaction between various parts of the body, the stage-by-stage nature of the entire process, and the numerous antagonistic, synergistic, and feedback relationships within the system, and between the system and the environment.

Of particular interest are analyses in which the non-learnt components of complex behaviour chains are further programmed and integrated by interaction with the environment. It has been found in several instances that animals raised without opportunity to learn a complex behaviour sequence to develop the elementary *components* of such an action chain, but have to have experience with the natural objects in order to mould them into an efficient skill. For instance, squirrels crack hazel nuts very expertly, by first manipulating them into a position in which they can gnaw selectively along the preformed groove where the shell is thin; then quickly weakening this groove, and finally cracking the shell along this weakened line. Squirrels raised alone and without nuts develop manipulating, gnawing, and biting, but their sequences are irregular and, because they gnaw all over the nut's surface, it takes them a long time to weaken the shell so that it can be cracked. Very long practice with nuts is required to develop this high-level skill (Eibl-Eibesfeldt, **14**).

Naturally, such analyses are only first steps, and while interference with processes inside the animal is required to further analyse the internal, 'maturational' aspects of development, further manipulation of the environment is used to determine which outside influences are effective and in what way they change the behaviour machinery. Thus it is important to know what exactly reinforces a response (for a recent summary of evidence see Hinde, **21**).

Both psychologists and ethologists have followed up this problem, though each in their own way, and both come to the conclusion that even where learning and other interactions with the environment occur, this learning is selective; for instance learning by sheer 'contiguity' of a response and any resulting stimulation is no longer believed to be a generally valid principle. Ethological evidence is mounting which shows that different species, even in roughly comparable situations, learn different things, and that different responses of the same animal are changed by different aspects of the environment. For instance, recognition of (i.e. selective responsiveness to) the correct conspecific sex partner is not conditioned in many species, whereas in Schutz's drakes it is (**48**). However, as we have seen, the *females* of these same species of ducks respond selectively to males of their own species even if raised with a different species. Numerous other examples of such 'predispositions to learn' are known.

This selectiveness of learning is also shown in such studies as Thorpe's, whose chaffinches do not acquire just any song they hear but such that conform in certain aspects to the song of the species. The concept of something like a pre-set 'expectancy', which makes the development of an animal at some stages particularly responsive to specific stimuli, is required, and is being introduced under a variety of terms, such as the 'innate teaching mechanism' (Lorenz, **41**), or a 'template' (Thorpe, **55**) against which the animal matches the incoming stimulation. However, the concept should certainly not be confined to *innate* teaching mechanisms—the template may itself be the result of previous learning processes. Thus further analyses of the way in which some birds acquire their song has shown that they may, by listening to experienced singers during an early critical phase, establish a template on the basis of this experience, to which they adjust their later song by comparing the sounds they make with it (Konishi, **30**).

Thus the conceptualisation in studies of behaviour development is evolving along lines strictly parallel to those in short-term causation studies. In addition, a process of interdisciplinary fusion is discernable in both fields. Lorenz began, in reacting against an overemphasis of the part played by external control, *by classifying*, stressing the fact that part of the total programming of an animal's behaviour machinery is done within the animal itself. By applying the term 'innate' to a class of behaviour patterns, however, he made it difficult, I believe, to move away from the preparatory, and at the time useful, classification of behaviours, and towards a programme of detailed analysis of the developmental processes. Yet it has been of great importance to realise that, by his own work and that of his pupils, he did call attention to the variety of patterns of interplay between internal and environmental programming. The analysis of the actual processes involved in internal programming has hardly begun, and so far many conclusions about such internal processes can only be supported by negative, eliminative evidence. But again, this evidence is not to be discarded lightly. For instance, while one could argue, as Lehrman (following Kuo, **31**) did, that the pecking movements of newly-hatched chicks could have been influenced by the head having been moved passively by the heartbeat while the chick was still in the egg, how could one possibly imagine a duckling having acquired, by any interaction with the environment, the peculiar movements of the wing that ducklings may make while fighting, plus the details of its orientation, which, when the behaviour first appears, is already 'calculated' to the size that the wing will not acquire until later? Facts such as those reported by Thomas and Schaller (**53**), that kittens raised without a mother and in complete darkness show the complicated 'hunting plays' the first time they are shown a dummy mouse; that sticklebacks reared without conspecifics show normal and selective fighting and courtship behaviour (Cullen, **86**); that young Blackbirds exclusively fed by white objects gape selectively at black dummies (Tinbergen and Kuenen, **61**); that male Chiffchaffs and Grasshopper warblers raised in isolation produce (unlike chaffinches) their normal, admittedly not very complex song (Heinroth and Heinroth, **19**)—such facts all point to important internal contributions to behaviour programming. On the other hand, the preoccupation (which I sense in workers such as Schneirla even now) with the possibility that the search for interactions with the environment at early stages will always show such interactions, prevents such workers from appreciating that not all the programming can possibly come from outside. Thus both the eliminative procedure and the restriction to manipulation of the environment fail to demonstrate, let alone analyse, the internal processes, as research of the type done by Sperry does, admittedly so far only for relatively simple levels. The true story of behaviour ontogeny will only be discovered by studying external and internal events and their interplay.

Of course these developments are of the greatest importance for an understanding of the ontogeny of our own behaviour. The experiments that are required can, for ethical reasons, never be done with human beings; deliberate interference with a child's development of the types and to the extent required is morally unacceptable. Clinical evidence is the best we can expect (see, for example, Bowlby, **4**); it should be collected on the largest possible scale and checked against the results of animal experiments. Ethologists claim that it is by no means proved, and is in fact highly unlikely, that manipulation of the environment (education in the widest sense) can mould Man's behaviour beyond the boundaries of innately determined ranges, although the extent of these ranges are hardly known. At the moment, it is neither scientific to claim actual knowledge of our innate behavioural equipment nor that we are infinitely mouldable—that, say, our aggression can be eliminated entirely by educational measures; such questions have to be considered undecided until they are properly investigated. The importance of modern work on animals lies at least partly in the fact that it is beginning to give us the tools required for such studies.

Unlike behaviour physiology and behaviour development, the study of behaviour *evolution* is still entirely in the hands of zoologists, neither psychologists nor physiologists having so far shown much interest in the field. Here ethology is widening its approach through increasing contact with ecology, genetics, and evolutionary studies in general—sciences developed by other biologists. The classical methods of genetics are now beginning to be applied to behaviour; correlations are being established between genetic, structural, and behavioural differences of different strains, populations, subspecies, and species; the inheritance of behavioural traits if followed through one, two, or more generations following cross-breeding, etc. The field, however, is still in its infancy, even though a large number of such correlations have been established at various levels of behavioural complexity. The gap between knowledge on the level of the genetic 'blueprint' and that of behaviour is still very large, and will only be filled by laborious analysis of behaviour machinery and its ontogeny.

At the same time, the part played by natural selection in moulding behaviour, as well as that of behaviour in creating new opportunities for natural selection, are being explored.

With regard to the first problem, several approaches are being adopted.

Natural selection is being applied artificially and its consequences studied. The grandest experiment done so far was the domestication of various animals, of which, however, the scientific documentation has been extremely sketchy. But refined measuring techniques and the use of fast-breeding animals such as the fruitfly *Drosophila* are now enabling us to apply selection pressures over a number of generations and to study its results within a time span of the order of a few years. The evolution of sexual isolation (i.e. lack of cross-breeding) between related strains—a phenomenon that is at the root of evolutionary divergence and is often a matter of mating behaviour—is an example. Species often split up into populations of slightly different constitution when they increase their geographical area. In the different corners of such an area populations go their own evolutionary way, partly because they are derived from slightly different, not quite representative samples of the heterogeneous original population, partly because, living in different environments, natural selection imposes additional differences. When such populations, in the course of their subsequent movements, come into contact again, their representatives may still cross-breed. But due to their different genetic makeup, the hybrids are usually less viable than both parent strains. In other words, natural selection discriminates from this moment against those members of the parent strains that cross-breed, by penalising their offspring more than pure offspring of either strain. Experiments are being done to test the hypothesis (derived from field observations) that such anti-hybrid selection favours, in both strains, the further development of mating preferentially within the strain. The results so far obtained are according to this hypothesis, that is, sexual isolation increased. Moreover this was shown to be due to certain evolutionary changes in mating behaviour (Crossley, **7**).

There is also an increasing number of studies (see, for example, Cain, **5**; Tinbergen, **59**) that are testing whether the behavioural characteristics of species are really such that they make each species singularly well-adapted to the ecological 'niche' in which it is living, and if so, which pressures of the environment it meets, how these pressures exert their influence, and finally how the animal deals with them. These studies are very similar to studies of survival value and the adaptedness of behaviour pursued in their own right, as mentioned above; the difference lies in their ultimate aim. Taxonomic and geographical studies can reveal the way in which populations have in the past expanded, split up, often subsequently overlapped, etc. When such studies are combined with ecological work, in particular investigations that test whether invasion of new habitats, of life in changing habitats, etc. produce changes in structure and behaviour of populations, one gets an idea of the extent to which natural selection must have been effective.

At the same time experimental studies are testing whether the behaviour differences between different species are actually adaptive, as they should be if selection has moulded them. This is done by comparing the success of a roughly normal population of a species with a population which in one of its characteristics deviates from the norm in the direction of another species. For instance, the several species of gulls differ in that while most species remove the empty eggshell after each chick has hatched some species leave the eggshell in the nest. It has been shown that eggshell removal is a corollary of the camouflage of the brood: eggs and chicks are protected by their colour against predators that hunt by sight, while the broken eggshell, which shows the white inside and edges, attracts predators, which then find and eat the brood (Tinbergen *et al.*, **60**). The species that do not remove the eggshell seem to have no need for it. For instance, one such species, the Kittiwake, nests on narrow ledges on vertical cliffs, where its broods are practically out of reach of predators.

The studies so far done along these lines begin to show how beautifully and intricately the behaviour of each species is adapted to its needs—within, of course, the limits of the overall capacities of animals at each evolutionary level. There are several circumstances that explain why our knowledge in this field is still meagre, and why the power of natural selection in moulding behaviour is still much underrated. First, few biologists are engaged in this type of research; the successes of the physical sciences having drawn much of the available talent to physiology, biochemistry, and biophysics. Second, our knowledge of the many pressures that the natural environment exerts on each species is still extremely poor. Third, the analytical method forces us to select for study one behaviour characteristic and one environmental pressure at a time, and one of the very natural reactions after one such study is to think that an animal could perform a particular task much better than it does. However, where this question has been studied in a broader context it has always become clear that there is, within the animal, competition between various activities, and that different pressures require different and not always compatible ways of meeting them. To mention a simple example: many young birds are camouflaged as a protection against certain predators. Camouflaged colour patterns are effective only when the animal is motionless. However these animals have to eat, and this requires motion. Thus their behaviour has to be a compromise; for instance, they will feed normally, and will 'freeze' when the parent spots a predator and calls the alarm. While they could feed more efficiently if they never had to freeze, and would be better protected against predators if they never had to move, they can do neither, and selection, rewarding overall success rather than any isolated characteristic, has produced compromises. A full understanding of the total problem of adaptedness therefore requires a synthesis of data concerning single behavioural traits and single environmental pressures.

The ways in which behaviour *creates* new opportunities for natural selection are also being studied. Perhaps the most basic behavioural contribution to evolution is the fact that every animal population continuously explores new habitats. Floating larvae of marine animals, and 'dispersal stages' of many land animals (for example spiders sailing on the wind with the aid of gossamer threads) are year after year carried in enormous masses outside their optimal habitat; young individuals of territorial animals are often repelled from already occupied habitats and forced to try to settle elsewhere. The majority of such 'pioneers' dies, but changing conditions in such newly explored habitats, or changes within the animals themselves, lead again and again to the invasion of previously unoccupied areas. Whenever this happens, the new environment inexorably enforces further evolution. The most famous example of this is provided by the South American finch-like birds that have invaded the Galapagos Islands, where they have developed into a group of widely divergent species (Lack, **32**).

Very spectacular examples of evolutionary change due to behaviour are found wherever the environmental pressure acting on an animal is exerted by another animal, such as in prey–predator relationships, or in social relations within the species, for example relations between sex partners or between parents and young. In relationships within a species, where success is often dependent on a fitting signalling system, it is usually difficult to say whether it is the signal that has caused the recipient to evolve a specific sensitivity to it or whether specific sensitivity of the recipient has enforced the development of a signal such as a movement, a sound, a scent gland, or a brightly coloured structure. But in some cases of intraspecific signalling this can be established. Wickler (**64**) has described a species of mouth-breeding fish, in which the male has, on its anal fin, a series of colour spots that are remarkably accurate two-dimensional 'pictures' of the eggs of the species. During mating, the female takes the eggs into her mouth immediately after laying, but for the eggs to be properly fertilised she has to snap up the male's sperm as well. She is made to do this by the male's displaying the egg-'lures'; in her attempts to snap these up she takes up the male's sperm. Clearly the evolution of the signal is an adaptation to the female's visual response to eggs, which in its turn is an adaptation to the eggs. Numerous examples are known in which behaviour of a predator forces the prey species to evolve defences, a simple case being the general motionlessness of camouflaged animals (movement being a powerful stimulus that would destroy the camouflage effect); more complex are the 'distraction displays' by which many ground-breeding birds lure predators away from their brood. In the struggle between predator and prey the prey species can also direct the evolution of the predator; thus various predators have developed lures that mimic the food of their prey species.

IV

In this admittedly sketchy outline of the development of ethology to date I may well have misjudged the relative importance of contributions made by this young science in comparison with those of sister disciplines. The exchange between these various sciences, in published work and, more importantly, in many personal contacts, has been so intensive and in addition so interwoven that it is impossible for any one person to keep a good or even an objective record. Moreover, many changes of viewpoint and of approach have occurred at an intuitive, unconscious level and have not been made explicit until at a late stage—they were expressions of general trends that were 'in the air'. It seems justified, however, to say that the behavioural sciences have in these sixty years moved gradually towards increased affinity with other natural sciences, and that ethology has contributed substantially to this development. In spite of its initial bold simplifications, its extreme positions in reaction against prevailing views, and its various other shortcomings, and in spite of the fact that the other behavioural sciences had moved independently towards both more scientific and broader methods of approach, I believe that it is fair to say that ethology has had a healthy, invigorating effect, not only on this process but also on hastening the fusion of many separate disciplines into one comprehensive main stream of truly biological research.

The future consequences of this development will undoubtedly be of great importance. The methods and concepts of this emerging biological science of behaviour are proving themselves so successful in animal studies that they will have to be applied to human behaviour as well; in fact such studies are already starting. Few will deny that there are many disturbing signs of malfunctioning of our own behaviour, particularly of our social behaviour. It is a task of the greatest urgency to try to find out how this came about. The most likely hypothesis is that the culturally determined changes in our environment (particularly our social environment) have outpaced adjustments in our behaviour; that genetic evolution is much too slow to achieve such adjustment; and that our individual behaviour is not sufficiently modifiable because we too are genetically restricted—the consequence of our still being adapted to the ancestral environment. Animal behaviourists agree that the conception of Man as an infinitely adjustable species, of which each individual can in its lifetime be modified behaviourally by educational measures to any desired extent, is not necessarily correct; in fact it is highly unlikely to be true. And even if it were largely true, it would still be abundantly clear that, in our ignorance of the natural control of behaviour ontogeny, we have not yet found the best ways of guiding the development of young individuals.

This consideration is relevant to a sketch of the growth of our science because further development may well be hampered by a kind of social negative feedback, a 'backlash'. As the scientific understanding, and through it the control of our own behaviour, will be seen as even a remote possibility, the resistance against this type of self-analysis may well increase. Whether such resistance can itself be eliminated by education is perhaps an even more basic problem, but even this is open to scientific investigation. Thus, inevitably, the developments I have tried to sketch will affect Man's attitude to himself.

REFERENCES

1 BEACH, F. A. (1955). 'The descent of instinct', *Psychol. Rev.*, **62,** 401–10.

2 BERKUN, M. M., M. L. KESSEN and N. E. MILLER (1952). 'Hunger-reducing effects of food by stomach fistula versus food by mouth measured by a consummatory response', *J. comp. physiol. Psychol.*, **45,** 550–4.

3 BLEST, A. D. (1957). 'The evolution of protective displays in the Saturnioidea and Sphingidae (Lepidoptera)', *Behaviour*, **11,** 257–309.

4 BOWLBY, J. (1960). 'Ethology and the development of object relations', *Int. J. Psycho-Analysis*, **41,** 313–17.

5 CAIN, A. J. (1964). 'The perfection of animals', *Viewpoints Biol.*, **3,** 37–63.

6 CRAIG, W. (1918). 'Appetites and aversions as constituents of instincts', *Biol. Bull.*, **34,** 91–107.

7 CROSSLEY, S. A. (1963). Doctor's thesis, Oxford University; see also Tinbergen, **59**.

8a CULLEN, E. (1957). 'Adaptations to cliff-nesting in the kittiwake', *Ibis*, **99,** 275–302.

8b —— (1961). 'The effect of isolation from the father on the behaviour of male Three-spined sticklebacks to models', *Tech. (Final) Rep. on Contract AF* 61(052)–29, *USAFRDC*.

9 DARWIN, C. (1859). *On the Origin of Species by Natural Selection*. London.

10 —— (1872). *The Expression of Emotions in Man and Animals*. London.

11 DESMEDT, J. E. and P. MONACO (1961). 'Mode of action of the efferent olivocochlear bundle on the inner ear', *Nature, Lond.*, **192,** 1263–5.

12 DETHIER, V. G. and D. BODENSTEIN (1958). 'Hunger in the blowfly', *Z. Tierpsychol.*, **15,** 129–40.

13 ECCLES, J. C. (1953). *The Neurophysiological Basis of Mind.* Oxford University Press.

14 EIBL-EIBESFELDT, I. (1963). 'Angeborenes und Erworbenes im Verhalten einiger Säuger', *Z. Tierpsychol.*, **20,** 705–54.

15 GROHMANN, J. (1939). 'Modifikation oder Funktionsreifung?', *Z. Tierpsychol.*, **3,** 132–44.

16 HARRIS, G. W., R. P. MICHAEL and P. P. SCOTT (1958). 'Neurological site of action of stilboestrol in eliciting sexual behaviour', *Ciba Foundation Symposium on the Neurological Basis of Behaviour*, London, pp. 236–51.

17 HASSENSTEIN, B. (1961). 'Wie sehen Insekten Bewegungen?', *Naturwissenschaften*, **48,** 207–14.

18 HEINROTH, O. (1911). 'Beiträge zur Biologie, namentlich Ethologie und Psychologie der Anatiden', *Verh. Vth. Int. orn. Congr.*, pp. 589–702.

19 —— and M. HEINROTH (1928). *Die Vögel Mitteleuropas.* Berlin.

20 HINDE, R. A. (1965). 'The integration of the reproductive behaviour of female canaries', in *Sex and Behaviour* (edited by F. A. Beach), pp. 381–416. New York.

21 —— (1966). *Animal Behaviour.* New York.

22 HOLST, E. VON (1939). 'Entwurf eines Systems der lokomotorischen Periodenbildungen bei Fischen', *Z. vergl. Physiol.*, **26,** 481–528.

23 —— and H. MITTELSTAEDT (1950). 'Das Reafferenzprinzip', *Naturwissenschaften*, **37,** 464–76.

24 —— and U. VON ST PAUL (1960). 'Vom Wirkungsgefüge der Triebe', *Naturwissenschaften*, **47,** 409–22.

25 HUXLEY, J. S. (1914). 'The courtship habits of the Great Crested grebe (*Prodiceps cristatus*); with an addition to the theory of sexual selection', *Proc. zool. Soc. Lond.*, 419–562.

26 —— (1923). 'Courtship activities in the Red-throated diver (*Colymbus stellatus* Pontopp.); together with a discussion on the evolution of courtship in birds', *J. Linn. Soc.*, **35,** 253–92.

27 —— (1963). 'Lorenzian ethology', *Z. Tierpsychol.*, **20,** 402–9.

28 JENNINGS, H. S. (1923). *The Behavior of Lower Organisms.* New York.

29 KNOLL, M. D. (1956). 'Ueber die Entwicklung einiger Funktionen im Auge des Grasfrosches', *Z. vergl. Physiol.*, **38,** 219–37.

30 KONISHI, M. (1965.) 'The role of auditory feedback in the control of vocalisation in the White-crowned sparrow', *Z. Tierpsychol.*, **22,** 770–83.

31 KUO, Z. Y. (1932). 'Ontogeny of embryonic behaviour in Aves. IV', *J. comp. Psychol.*, **14,** 109–21.

32 LACK, D. (1947). *Darwin's Finches.* London.

33 LEHRMAN, D. S. (1953). 'A critique of Konrad Lorenz's theory of instinctive behaviour', *Q. Rev. Biol.*, **28,** 337–63.

34 —— (1961). 'Gonadal hormones and parental behaviour in birds and infrahuman mammals', In *Sex and Internal Secretions*, ed. W. C. Young. Baltimore.

35 LISSMANN, H. W. (1946). 'The neurological basis of the locomotory rhythm in the spinal Dogfish (*Scyllium canicula, Acanthias vulgaris*). II', *J. exp. Biol.*, **23,** 162–76.

36 LORENZ, K. (1927). 'Beobachtungen an Dohlen', *J. Orn., Lpz.*, **75,** 511–19.

37 —— (1931). 'Beiträge zur Ethologie sozialer Corviden', *J. Orn., Lpz.*, **79,** 67–120.

38 —— (1935). 'Der Kumpan in der Umwelt des Vogels', *J. Orn., Lpz.*, **83,** 137–213; 289–413.

39 —— (1937). 'Ueber die Bildung des Instinktbegriffs', *Naturwissenschaften*, **25,** 289–300; 307–18; 324–31.

40 —— (1939). 'Vergleichende Verhaltensforschung', *Zool. Anz.* Suppl., **12,** 69–102.

41 —— (1965). *Evolution and Modification of Behavior*. Chicago.

42 MORGAN, L. (1896). *Habit and Instinct*. London.

43 PAVLOV, J. P. (1926). *Die höchste Nerventätigkeit (Das Verhalten) von Tieren*. München.

44 PERDECK, A. C. (1964). 'An experiment on the ending of autumn migration in starlings', *Ardea*, **52,** 133–40.

45 ROEDER, K. C. (1962). 'Neural mechanisms of animal behavior', *Am. Zool.*, **2,** 105–15.

46 SCHNEIRLA, T. C. (1966). 'Behavioral development and comparative psychology', *Q. Rev. Biol.*, **41,** 283–302.

47 SCHNEIRLA, T. C., J. S. ROSENBLATT and E. TOBACH (1963). 'Maternal behavior in the cat', *Maternal Behavior in Mammals*, ed. H. Rheingold. New York.

48 SCHUTZ, F. (1965). 'Sexuelle Prägung bei Anatiden', *Z. Tierpsychol.*, **22,** 50–103.

49 SEVENSTER-BOL, A. C. A. (1962). 'On the causation of drive-reduction after a consummatory act', *Archs néerl. Zool.*, **15,** 175–236.

50 SHERRINGTON, C. S. (1906). *Integrative Action of the Nervous System*. London.

51 SPERRY, R. W. (1959). 'The growth of nerve circuits', *Sci. Amer.*, **201,** 68–76.

52 SUTHERLAND, N. S. (1962). 'The methods and findings of experiments on the visual discrimination of shape by animals', *Exp. psychol. Soc. Monogr.*, 1.

53 THOMAS, E. and F. SCHALLER (1954). 'Das Spiel der optisch isolierten, jungen Kaspar-Hauser-Katze', *Naturwissenschaften*, **41,** 557–8.

54 THORPE, W. H. (1961). *Bird Song*. London.

55 —— (1963). *Learning and Instinct in Animals*. London.

56 TINBERGEN, N. (1953). *Social Behaviour in Animals*. London.

57 —— (1964). 'Aggression and fear in the normal sexual behaviour of some animals', in *The Pathology and Treatment of Sexual Deviations*, ed. J. Rosen, pp. 3–23. Oxford University Press.

58 —— (1965). 'Some recent studies of the evolution of sexual behaviour', *Sex and Behavior*, ed. F. A. Beach, pp. 1–34.

59 —— (1965). 'Behavior and natural selection', in *Ideas in modern biology*, ed. J. A. Moore, pp. 521–42. New York.

60 —— *et al.* (1962). 'Egg-shell removal by the Black-headed gull *Larus ridibundus* L.; a behavioural component of camouflage', *Behaviour*, **19,** 74–118.

61 —— and D. J. KUENEN (1939). 'Ueber die auslösenden und die richtunggebenden Reizsituationen der Sperrbewegung von jungen Drosseln', *Z. Tierpsychol.*, **3,** 37–60.

62 VERWEY, J. (1930). 'Die Paarungsbiologie des Fischreihers', *Zool. Jahrb. Allg. Zool. Physiol.*, **48,** 1–120.

63 WATERMAN, T. H., G. A. G. WIERSMA and B. M. H. BUSH (1964). 'Afferent visual responses in the optic nerve of the crab', *Podophthalmus. J. Comp. Cell. Physiol.*, **63,** 133–55.

64 WICKLER, W. (1962). 'Zur Stammesgeschichte funktionell korrelierter Organ- und Verhaltensmerkmale', *Z. Tierpsychol.*, **19,** 29–64.

65 WHITMAN, C. O. (1919). *The Behavior of Pigeons*. Carnegie Institute, Washington, Publication 257, Vol. 3, pp. 1–161.

66 WIEPKEMA, P. R. (1961). 'An ethological analysis of the reproductive behaviour of the Bitterling', *Archs néerl. Zool.*, **14,** 103–99.

(*From the Department of Zoology of Oxford University*)

16
The Search for Animal Roots of Human Behaviour[1]

It is often said that Man is unique among animals. It is worth looking at this term 'unique' before we discuss our subject proper. The word may in this context have two slightly different meanings. It may mean: Man is strikingly different—he is not identical with any animal. This is of course true. But it is true also of all other animals: each species, even each individual is unique in this sense. But the term is also often used in a more absolute sense: Man is so different, so 'essentially' different (whatever that means) that the gap between him and animals cannot possibly be bridged—he is something altogether new. Used in this absolute sense the term is scientifically meaningless. Its use also reveals and may reinforce conceit, and it leads to complacency and defeatism because it assumes that it will be futile even to search for animal roots. It is prejudging the issue, for it is not a conclusion based on careful analysis, rather it may be a hasty judgment, almost a premise.

I submit that it will be useful, even imperative that we use the word in its first, relative sense, that of just being strikingly different. On the one hand we recognise that Man is unique, but on the other hand we also recognise that he is an animal—of a kind. One is greatly tempted to use an Orwellian phrase and say no more than: 'all animals are unique, but Man is more unique than others'.

The aim of my talk will be to explore how we can find out how unique Man really is. And its keynote will be to argue that in spite of much that we know, or suspect, our real knowledge is really extremely poor. And I have to stress in advance that I shall do no more than extend some tentative feelers, no more than groping my way in this very difficult subject.

Let us first limit our field. The uniqueness of Man is not a matter of

[1] (Unpublished lecture given in a series 'Social Studies and Biology' at Oxford University on 27 October 1964.)

his structure. His body and its functions are in general very similar to those of other mammals—that is why medicine can study in animals the functions of, say, kidneys, of the heart, of eye and ear, even the basic functions of the nerve cells, and extrapolate with confidence.

The uniqueness of Man is agreed to be a matter of behaviour. If it is to be reduced to structural characters, it is the brain that is unique, that functions in a unique way. There are of course links with other structural characteristics, but attempts (made in the past) to reduce Man's behavioural uniqueness to, say, the possession of hands, or to his upright posture, have not been very convincing—they have been too simplistic, too one-sided.

The way the biologist approaches this problem is based on his knowledge of the fact of evolution. No informed person can doubt any more that Man has evolved, slowly and very gradually, from ancestors which were far more similar to other mammals than Man is now. This means that everything Man is and does now must have evolved, through a long series of minute evolutionary steps, from what his animal ancestors were and did. Man has diverged very gradually from monkey or ape-like stock to what he is now, just as modern closely related animal species have diverged from common stock.

How do we trace this process of gradual divergent evolution? It is necessary to realise that we have to try to reconstruct a series of processes each of which was, in its particularity, unique. Each of these steps happened once in the past, and each of these events of the past is for ever lost to direct observation, unlike contemporary and future evolution. But biologists have evolved indirect methods which allow us to sketch out the most likely way in which evolution must have happened. As far as structural properties are concerned, we use a kind of historical document; fossils—which we can date and which therefore we can place on a time-scale. This procedure has allowed us for instance to say with confidence that whales have developed from land mammals, that bats have reconstructed their forelimbs so that they became wings. Lobsters have enormous claws which are used for the seizing and crushing of food, but lobster's ancestors used these appendages as their first walking legs. They have been secondarily adapted to a new function. Incidentally: this example highlights a relevant point about the concept 'new', in the sense I used it when I said: 'Man is something altogether new'. No evolutionary step has ever produced something really new, but merely changes in something existing. When we, subjectively, call a character 'new' we mean no more than that it is so different from what we see in other animals that the differences strike us more than the similarities.

But to return: in tracing the evolutionary past we also use a second, less direct method. Even if we had no fossils of ancestral whales we would still know that they must have been land mammals originally. This we conclude from comparison. Two such comparisons are made. (1) Whales share the majority of their characters with mammals, in spite of their superficial similarity to fish: and (2) because the great majority of mammals are land forms, because the whales are so aberrant, we conclude that whales descend from land mammals, and not the other way round. A similar argument applies to bats, and to lobsters—in fact, throughout Comparative Biology.

Because the comparative method gives on the whole the same results as the study of fossils (i.e. the comparative method has been 'gauged' against the more reliable palaeontological method), it is often used alone, mainly in those cases where fossil evidence is poor or absent. Now fossils don't behave, and for tracing behavioural evolution we have to rely entirely on comparison. This works on the whole well—for instance in those cases where we have classified species as being closely related on the basis of (1) similarities of behaviour, (2) similarities of structure, and (3) fossil evidence, we find on the whole conformity of results. This means that our method for tracing animal roots of human behaviour must be comparison. We simply have no better method.

Of course this comparison works best when the differences between the species compared are small. Since Man is undoubtedly very different even from its closest relatives, our task is a very difficult one.

Unfortunately there are more difficulties. Similarities between different species can evolve in two entirely different ways. In closely related species they often point to slight evolutionary divergence from common ancestral characters. For instance, the feet of horses, rather different from, yet similar to the feet of other mammals, have undoubtedly diverged from ordinary, five-toed mammalian feet. Wings of bats have diverged from walking forelimbs. But the wings of bats are also very similar to the wings of some extinct flying reptiles—but they have evolved independently, they have 'converged', in common adaptation to a similar function. Or, to mention just one other example, such organs as gills and lungs have developed quite independently, convergently, in many very different animal groups.

Similarly, species-specific behaviour patterns have evolved convergently in many cases. This is true, for instance of swimming in whales, in fish and in some dragonfly larvae. When therefore I am going to compare behaviour patterns in Man with those of other animals, I may not be dealing with similarities that reveal descent from common stock—I may have to do with convergencies, which means that similarities may be superficial. This danger is larger the less closely related Man and the animals considered are.

These are not even all our difficulties. By far the most serious obstacle in behaviour comparison is the fact that we describe and interpret our own behaviour in terms quite different from those we use to describe animal behaviour. We describe human behaviour really in three kinds of terms:

1. We may really describe what movements a man makes. This is what we do (though in a superficial way) when we say, e.g. 'He ran as fast as he could'; 'She smiled at him'.

2. We often describe, not movements, but a subjective experience: 'He became very angry'; 'He fell in love'.

3. We also describe behaviour in terms of effects of movements: 'He ran for cover'; 'His overriding aim was to silence opposition'. As a rule we even mix these types of descriptions ('she irritated him by smiling indulgently'), and one of our difficult tasks is to extract from such statements the objective, descriptive part which refers to observed movements and distinguish them from interpretation of function or subjective phenomena.

How when we describe animal behaviour, we try to use exclusively the first method, that of describing actually observed movements. This is because whatever subjective experiences animals have (and even though most of us *believe* they must have them) we have no scientific means of deciding whether or not they experience them. Nor do we know to what extent the future effect of an animal's behaviour really controls at any given moment what it is going to do.

If we want to compare human and animal behaviour, the first thing we have to agree on is the use of a common language. But, as I said, we are used to describing human behaviour in a mixture of terms, some objectively-descriptive, some relating to subjective experiences, others again relating to the 'aim', the effects of our behaviour. This discrepancy of method, the use of entirely different languages in descriptions of human and of animal behaviour, is really at the root of our lack of comparative knowledge.

So, while we have to agree that comparison is the only method at our disposal, such comparison is hampered by difficulties of four kinds:

1. The method is itself indirect.

2. The gap between Man and animals is so large that it is often difficult even to guess at 'animal roots'.

3. We have to distinguish between true similarities (which are due to common descent) and convergencies (which are not).
4. We have not yet developed a disciplined use of language in describing human behaviour.

And this makes it so extremely difficult to find even the starting point for our inquiry: to state which are those traits of behaviour in Man that are typical of our species. Which behaviour shall I single out for discussion?

The best one can do is to select some aspects of human behaviour on which there is at least a certain concensus of opinion. But at the same time, in view of the deplorable state of this comparative ethology, it would be futile to discuss human traits that seem as yet so far removed from animal behaviour that we have not even a hunch about their animal roots. I have therefore selected some problems that may strike you as pedestrian and as being far removed from the more challenging problems. My aim, however, is not, in the first place, to demonstrate the existence of animal roots but to discuss some points of *method*—to discuss how we *could*, and should proceed.

I shall start with a relatively straightforward issue. It has often been pointed out that Man, himself a product of evolution of a type similar to that which has created all other animal forms, namely adaptive hereditary change, has now embarked on a new type of evolution, which Huxley calls 'psycho-social evolution'. I prefer the term 'cultural evolution'. It is based on *accumulated* transfer, by tradition, from one generation to the next, of knowledge (or *phenotypic*) behaviour changes, i.e. changes acquired through individual experience. Our culture is very different from that of Cro-Magnon Man, but genetically we may not have changed much—most of our modern attributes are due to the accumulation of transferred knowledge. We differ from animals not merely in the extent of what we can ourselves learn, but in the progressive (and steadily accelerating) accumulation of experience through the generations. How unique are we in this respect?

However little we have studied this in animals, we know that at least our nearest relatives, apes and monkeys, do transfer newly acquired behaviour. Chimpanzees on the whole do not know how to open coconuts. Elder has found that some individuals crack them by smashing them against a hard substrate. He also showed that if individuals who don't know how to open them see others crack the nuts, many of them quickly learn how to do it. Similar observations have been reported from Macaque monkeys in Japan, and our Japanese colleagues use the terms 'acculturation' for this phenomenon. Even more interesting: individuals are not willing to learn just from anybody, infants learn from mothers more than the opposite, animals low in the pecking order learn from their 'superiors' rather than the other way round. Approaching the subject from the other side: to what extent do *we* really learn from others? Is our willingness to learn always the same? How monkey-like are we perhaps in this respect? Schoolboys have to have respect for their teachers or they will resist being taught. Studies of the history of languages have shown that new developments in language and in other aspect of culture radiate out from centres of power—large towns in medieval times, admired nations in modern times.

Considering the very primitive stage of our knowledge of the behaviour of our fellow-Primates (some insects, fish and birds are far better known) and taking into account how difficult it is to get reliable observations on such cultural transfers, we cannot judge the extent of 'psycho-social' evolution in Primates other than Man, but once we are aware of its occurrence we might well discover more nearly human types of transfer from one generation to the other—and also more primitive, non-rational determinants of learning in ourselves.

The phenomenon is certainly not confined to Primates; simple examples of transfer are known of animals of a much lower behavioural level. Some song-birds, for instance, learn the song which is typical of their species by listening to the song of their parents. There is an amusing and well-documented case of a male bullfinch which, because it was raised by a pair of canaries, acquired the full canary song. This bullfinch was transferred to another place out of earshot of canaries well before it raised young itself. There it mated with a female of its species, and raised young with her. The sons of this bullfinch sang like canaries. One of them raised young of his own in his turn, and the sons of this bird sang like canaries too, even though neither they nor their father had ever met a canary. One could call this a primitive kind of acculturation. But in animals this has never yet been shown to be accumulative.

Geese live in families for almost a year; these families do not break up until the parents start a new brood in the subsequent year. Parents call their young away from places where they (the parents) have seen a fox on previous occasions. In K. Lorenz's institute successive generations of geese kept luring their young away from places where their ancestors had seen foxes, and this they kept up years after the erection of a fox-proof fence which prevented any fox from visiting these dreaded sites.

In this connection it is worth bringing up the question to what extent Man's tendency to learn is really different from that of animals. We know, of course, that animals learn a great deal. But it is less well known that much of this learning is far from a passive affair. Many animals actively set out to learn; they have special behaviour patterns of which the exclusive function is to create the opportunity to learn certain things. They are 'curious', that is (in biological language) they approach certain objects, situations or other animals in such a way that they can get all the sensory information about them that is possible—in short they set out to explore. Once the information is acquired they switch to other behaviour. Rats explore, and by doing so familiarise themselves with any change in a familiar environment; prospective sexual partners and social companions meet and investigate each other, and on the basis of information received 'decide' on their next course of action. Many bees and wasps make special 'locality studies' upon leaving their nest or burrow, and it has been shown that it is during these locality studies that they actually learn the location of their base in relation to landmarks which they later use for homing.

In some such animals the knowledge acquired through exploration is passed on to companions or young—but again, this 'cultural transfer' is not accumulative.

Looking at our problem from the other side, we may well ask how much of our exploratory behaviour is a really novel character? To what extent is even our highest form of exploration, scientific inquiry, an elaboration of more primitive forms of 'curiosity'? We all know that the tendency to explore diminishes with age, and that, while our brain machinery used in exploration may still be good, we do less exploring as we grow up, simply because the urge to explore wanes.

The whole issue of the extent to which our social environment moulds our behaviour, and to what extent genetical differences steer our psychosocial evolution, needs reconsidering. Many of you will have followed the correspondence in *The Observer* following the Newsom letter, in which qualitatively different education for women was proposed. Newsom's proposal was partly based on the consideration that society might well require different contributions from women than from men, and partly on the supposition that men and women are inherently, that is genetically, different, not only in their structure, but also in their behaviour, including their inclinations to learn; girls might well have inherently different interests than boys. Of course if the latter were true, education of boys and girls should be different if we want to get the best out of both, and if both are to grow up to happy and fulfilled members of society.

The opinions of correspondents differed widely. The disturbing thing about this correspondence was that premises were accepted but not challenged; both sides based their confident assertions on intuitive or even emotional attitudes. What we really need to know is to what extent and in what ways male and female infants do have different inherent interests. This is a straightforward enough question, one that can be solved in animals by raising the sexes under identical circumstances and testing whether their learning takes different courses. This is an experiment which for ethical reasons is difficult to do in human beings, but a great deal of clinical evidence points to the conclusion that boys and girls *are* different in these respects.

My point is that here is an extremely important issue, one on which far-reaching decisions may well be taken in the near future, and which should be decided on the basis of scientific information of a kind which we do not yet have, but which we can acquire. Yet too few people draw the main conclusion: that we have to *study* this aspect of human behaviour before we try to *control* it.

Let us now turn to another subject. Many people consider Man unique in his ability to grasp cause-effect relationships, whereas animals are considered to be, so to speak, at the mercy of stimuli, internal and external. It is worth considering whether a primitive grasp of cause-effect relations does not exist in animals, and also what exactly makes Man draw the conclusion that there is a relation between cause and effect.

If, for a moment, we leave very sophisticated scientific causal research out of consideration, the general evidence on which, in everyday life, we draw this conclusion, is roughly of this kind: when A always precedes B, and B does not occur without A having occurred, then we conclude that A is the cause of B. In science we arm ourselves against being blinded by occasional coincidences by observing sequences of events repeatedly. But in everyday life we are not so cautious: very few A–B sequences, or even one, makes us 'jump to conclusions'.

Just one anecdote: I once observed, together with my friend Konrad Lorenz, my 3-year-old daughter when she was trying to put her little scooter against the wall. On her first attempt it toppled over. The same happened on the second, and the same again on the third occasion. Then she did it once more, and this time, having lost her temper she gave the scooter an angry kick. And it stood! I wish you could have seen the triumphant expression on her face: *the kick had done it*! I could not help remarking to Lorenz: 'Look! she is as stupid as . . . as an adult'. I trust you see the point I want to make.

Some of you may remember with me how many people blamed the radio for the (alleged) deterioration of the weather; how later TV was blamed, and still later the atom bomb. All these conclusions, justified or not, were at best based on what amounts to *one* A–B sequence—a coincidence, usually selectively remembered.

Even in scientific inquiry we rely basically on observation of A–B sequences but we take care to repeat such observations under controlled conditions and so sorting out mere coincidences.

Now among animals, responses to A on the assumption that B is to follow are of course well known, because this is what happens as a consequence of Pavlovian conditioning; if, say, a bell is rung a number of times just before a dog is fed, the dog will soon respond to the ringing of the bell (the 'conditioned stimulus') as it would to stimuli emanating from the food itself. It acts in the same way as a man who has learned, through repeated observation of A–B sequences, that B is the effect of A, or, to put it the other way round, that A is the cause of B. Can we compare this with our 'insight' in cause-effect relationships? It seems quite possible that here again our use of different languages obscures a basic similarity; if we would describe our own behaviour (or, to make things easier, that of our fellow humans) in the same terms as that of the dog (instead of using terms such as 'knowledge', 'insight' and other terms that refer really to our subjective experiences) we might not feel that the gap is so large. It is at any rate desirable to ask here too: to what extent is our behaviour really novel, to what extent an elaboration of animal behaviour? I am not saying that the link will be easily found—I am merely stressing the need for truly comparative studies, and of an appropriate use of language.

Another rather widely accepted difference between Man and animals could be expressed, again sketchily, in the following terms: animals react 'blindly', slavishly, to internal and external conditions, stimuli, which immediately precede its behaviour—but Man 'reasons', 'acts deliberately'. Again, can we translate this term 'reasoning' into something which we can study in animals? We all know from introspection that something like this is involved: we can, before we act, imagine, think out, what the probable effects will be of one or another of several contemplated courses of action. We need not really act and then see what happens—we can internally imagine the effect, and the *expected* effect may control our decision as to what to do. The machinery which allows us to do this is unknown, but that we do it is a fact. This term 'expectation' is again a subjective term. Can we translate it into behaviour that we can observe?

The first steps in this direction have been taken, and I shall try to explain a few observations which I think are relevant.

Many insects show what is known as an optomotor response.

When they are at rest, and their environment is made to move (for instance when they are inside a hollow cylinder painted on the inside with vertical black and white stripes, and this cylinder is turned) they will turn with it, and keep their position constant with respect to their visual environment. This response allows, for instance, flying insects such as hoverflies to stay where they are even in strong wind. The stimulus that makes them do this is the movement of the projection of the environment over the sensitive retina of the eye. Now when such an insect is in a stable environment, and decides to move spontaneously, the effect is of course a movement over the retina in a reverse direction. The optomotor reaction 'ought' to make them turn back at once. But it does not; the insect is free to move in spite of the consequences registered on the retina. This used to be explained by saying that, when the insect moves spontaneously, the optomotor response is suppressed. Two German zoologists discovered that this is not true. They turned a fly's head over 180 degrees along the longitudinal axis of body, and fixed it that way (a fly can itself easily twist its head to that extent without harm). Now, of course, the movement over the retina will be the exact reverse of that normally experienced. Now whenever such an insect moved spontaneously it went into a mad spin. It did show an optomotor response, but in a direction opposite to the one required to reduce movement on the retina. Of course, by doing this it made matters worse. This proved that the optomotor response was not suppressed. But what then allowed the normal insect to move spontaneously at all? The only possible conclusion was that when it moved spontaneously it did expect movement over the retina, but ignored it as long as it occurred in the expected direction and at the expected rate. The only difference between the stationary and the spontaneously moving fly was that the stationary fly was 'set' not to move, to correct its position, when the retina image was stationary, and the spontaneously moving fly was 'set' not to correct its position, or rather its movement, as long as the report coming back from the environment was 'movement over the retina of a certain kind'. One could express this perfectly well, as I have done, by saying: the animal 'expected' a certain type of effect. But in neutral terms one could also say: it kept doing nothing, or kept turning as it wished (i.e. did not correct), as long as the 'feedback' was in accordance with what its nervous system was 'set' for. How this 'setting' is done is still a matter of conjecture. but that it is done is obvious.

This may seem very far removed from our own expectations—ours are more complex, and they take us much further ahead in time. But perhaps the following fact will be more appealing.

A mother hen can make a chick peck at the ground and pick up food by making pecking movements in front of it. This, even when done in front of a chick by a cardboard dummy of a hen's head, does make the chick peck. Now the Cambridge zoologist Turner trained a chick to find food in a concealed corner A of its pen. In the subsequent experiment a dummy of a hen's head was made to make pecking movements in another corner, B. The dummy had two kinds of grain in front of it, brown and green grains. When it pecked at the brown grain, the chick would run to corner A, and there select the brown grains from a mixture. It was taught to 'expect' brown grain in corner A. But one could equally well express this by saying that the stimuli from the mother 'set' it to run to corner A until it had seen brown grain, and then respond to this by eating. Something had happened in the chick, an internal process, that made it act with reference to the future, 'expected' effect of its behaviour. This is by no means an isolated case—many similar types of behaviour are known. Here again, there is no way of saying whether or not this could be the root of expectation in ourselves. Again I have to make the point that good comparisons will not be possible until we describe human and animal behaviour in the same terms.

As before, we can look at the problem of 'reasoned' versus 'forced' behaviour ('slavish' responses to external and internal conditions) from the other side, and ask how much of our own behaviour is really controlled by our coolly judging the probable future effects? Our animal roots can perhaps be seen best in cases where people act 'against better judgment'—when they do something which they very well know will lead to an unsatisfactory result. When man is subject to a very strong non-rational urge such as hunger, or a sexual urge, he may do many things against his better judgment. As an inmate of a German prison camp I have, under conditions of extreme hunger, seen men of normally high moral standards steal food from starving fellow prisoners. And we need only ask criminologists how many crimes, with disastrous results for the criminal, are committed under the influence of the sexual urge—not merely sex in its crudest sense, but also in the form of an uncontrollable desire to show off to a female.

All this may be commonplace—but is the significance of such facts fully realised?

Perhaps the most dangerous of these non-rational urges is our aggressive tendency, and it may be useful to look at this behaviour system a little more closely. Aggression is often looked upon, and from a point of view of management of human affairs rightly so, as an anti-social, a disruptive trait. But in animals this is only half the story. Aggression (of many different types) is deeply rooted in many animals. There are also many indications that it is in many species not learnt, but 'innate' in the sense that it develops normally in animals that have grown up without any opportunity to learn it in any of the conventional ways. Studies of the biological functions of aggression, of the way it affects success, begin to show that it is always part of a species' adaptive equipment, that it is definitely useful. For instance it is through aggression (*and its concomitant*, *fear*, *in the attacked individual*) that many animals space themselves out and prevent overcrowding. This is often (though by no means always) linked with the attachment to a particular area; such a defended area is called a territory. In most territorial species a territory is defended by individuals, or by breeding pairs, but there are also species, such as wolves and their relatives, and some Primates, where groups or clans defend a group territory.

Now aggression in animal species rarely leads to killing, or even to harm. This is partly because each species has a delicately balanced attack-and-escape behaviour; an animal will often retreat or keep its distance in response to attack or threat rather than fight back. But they also have another safeguard: a beaten or really cornered animal adopts a special posture (the submissive or appeasement posture) which has the curious effect of directly inhibiting the aggressive behaviour of the attacker—of soothing or appeasing him.

The comparative ethologist is not prepared to believe, without proper investigation, that Man is not inherently aggressive; he is not convinced that he can, through suitable education, be moulded into a non-aggressive individual. He thinks that we should investigate this very fully. In the absence of sufficient facts, he is inclined to think that Man may well be originally similar to some of his nearest relatives, and may well have a deep-rooted, irrational tendency to defend group territories, and also other tendencies to agonistic interaction, such as sexual rivalry, and rivalry over rank or status.

To know how deeply rooted are the aggressive behaviours of Man, is, of course, of the greatest importance when we want to control them and render their effects harmless. In inherently aggressive animals suppression can have temporary results, but it may in the long run lead to a kind of accumulation, and to terrible explosions of ruthless aggression, which makes animals fight themselves to death, or kill their own mates and young.

Biologists have discovered ways apart from appeasement in which animals render aggression harmless. An aggressive animal faced with a superior opponent will often redirect his attack—not merely towards weaker individuals (as we tend to do when we have been told off by our boss and take it out on wife or children) but even towards lifeless objects, such as sticks and leaves. It might be worth exploring to what extent we can re-channel our aggression, and even sublimate it, for instance by making concerted 'attacks' on nature—by 'conquering' the sea, as the Dutch do, or by 'conquering' space, or pollution, or disease.

Man's aggression has become so dangerous because inhibition by fear is deliberately reduced by whipping-up group aggression and by various training methods, also by instilling even greater fear for the consequences of lack of aggression, all of which is facilitated by our possession of means of mass communication. It has also been said that Man, by developing long-range weapons, has rendered the appeasement behaviour of his fellow-men ineffective because long-range appeasement has not kept track with long-range aggression. It is easier to drop an atom bomb from a plane, or to send off a rocket, than to strangle a man or a child by one's own hands—one does not actually see the misery one causes.

I dealt with aggression in more detail because the problem concerns us vitally. Yet what are we really doing about its study? Why are we not putting more intensive effort in a large-scale scientific study of aggression? Why don't we even try to apply what little knowledge we have?

Investigation of this human, and yet so animal, behaviour should proceed over a very broad front indeed. Our ignorance is really appalling. It has for instance been suggested by Lorenz, that even if we would succeed in making non-aggressive men, we might not improve matters at all. Aggression, especially group aggression, has in the animal kingdom a correlate of very positive value: true friendship seems to develop particularly between members of a group who are jointly aggressive to an outside group. To what extent are aggression and the inclination to from such social ties linked? We have no idea—but the problem is open to investigation. My own guess is that Lorenz is rashly jumping from an aspect of genetic evolution to one of phenotypic control.

I hope that you have not misunderstood this admittedly rather pedestrian talk. I have not touched on what many of us may well consider Man's highest behavioural attributes—speech, or sense of beauty; ethics; or religion. This is not because I believe that these have no animal roots. I did not discuss them because our ideas about their possible origin are even more speculative than the ideas I have been discussing, even though plausible suggestions have been made with respect to all these higher achievements. The principal aim of my talk has been to emphasise the really staggering extent of our ignorance, and at the same time to show that Biology can offer promising *methods* for a truly comparative biological study of human behaviour. If we want to understand our animal heritage—which *must* be there because we have descended from animals—we shall have to apply the same methods in human ethology as are beginning to be applied with success in animal ethology. I am convinced that our disinclination to do so has to do with deep-rooted aversions against subjecting ourselves to scientific study, aversions of which we may not be conscious. Will all respect for what has already been done, we have to admit that the biological study of Man's behaviour is as yet no more than a programme. But it is a feasible, and also an urgent programme.

(*From the Department of Zoology of Oxford University*)

17

Early Childhood Autism—an Ethological Approach[1] (1972)

The psychic life of the baby must be observed as Fabre observed his insects, going in search of them to surprise them in their natural environments, and lying hidden so as not to disturb them.

Maria Montessori, early twentieth century Quoted from: (1936) *The Secret of Childhood.* Edition Orient Longmans, Bombay. p. 45

It is not informative to study variations of behaviour unless we know beforehand the norm from which the variants depart.

Peter B. Medawar (1967) *The Art of the Soluble.* Methuen, London. p. 109

Introduction

The condition now widely called 'early childhood autism' was first identified and described by L. Kanner in 1943 (**17**). While there is no full agreement among the experts about the diagnostic criteria, the nine points originally listed by the working party under Creak (**7**) on what was then called the 'schizophrenic syndrome in childhood', and later slightly modified by O'Gorman (**19**), are now fairly generally accepted. As the term autism implies, the patients appear to be withdrawn, and in particular fail to establish social bonds; periods of underactivity alternate with hyperactive periods; when active the children often repeat simple, seemingly senseless 'stereotypies'; they often fail to respond to sensory stimuli but are on other occasions hypersensitive; learning is generally retarded and in particular speech either does not develop or regresses; and they often throw 'temper tantrums'.

Whereas in Kanner's time autism seemed to be rare, there are now, in Britain alone, some 6,000 children officially diagnosed as autistic, and although part of this increase may be merely due to better

[1] Condensed version of E. A. Tinbergen and N. Tinbergen (1972). *Early Childhood Autism—An Ethological Approach*, in *Advances in Ethology* (Supplements to *Z. Tierpsychol.*), **10,** 1–53. Parey, Berlin.

recognition of the symptoms, there may very well be a real increase of incidence.

Autistic children give the impression of being desperately unhappy, as are their parents; and although a proportion of autists recover, many of them end up as severely disturbed adults. Yet there is a widespread suspicion that many autists may well be of above-average potential, and that recovery is often so to speak 'just round the corner'. An increasing amount of effort is therefore going into attempts to understand the causation of autism, but no agreement has so far been reached—in fact opinions differ widely. Some authors are convinced that the deviation is either genetic or organic, and at any rate not due to upbringing (Wing, **30**); others suspect environmental causes (Bettelheim, **1**); some authors believe that the syndrome is primarily the result of overall overarousal (S. J. Hutt *et al.*, **14**; Hutt and Hutt, **13**) or of too much overall sensory input (Stroh and Buick, **24**); others again consider the speech defects as primary (Rutter *et al.*, **23**). Not astonishingly, the phenomenon is still considered as baffling; and equally understandably, therapies differ widely in their aims, their methods, and their success.

Although we are not ourselves professional students of autism, we have been interested in the problem ever since, years ago, Drs S. J. and C. Hutt consulted us about their attempts to study autism by means of ethological methods. Our interest remained more or less dormant until, in 1970, they wrote (in Hutt and Hutt, **16**, p. 147) that '. . . apart from aversion of the face, all other components of the social encounters of these autistic children are those shown by normal non-autistic children'. Since we knew face aversion well from normal children, this remark prompted us to reconsider certain of our observations on normal and autistic children, and to apply to these observations certain analytic procedures which had for years been used with success in the study of social interactions in higher animals. This seemed to make sense since autistic children hardly or never speak, and so force one to look at their non-verbal behaviour in much the same way as at the non-verbal 'expressions of the emotions' of animals.

Looking back at our knowledge, accumulated over the years, of the behaviour of normal children, and reconsidering what we knew of autists, both from incidental observations and from a study of the literature, we realised that the situations in which normal children could show components, and sometimes even the whole of Kanner's syndrome, could give us a clue about the causation of autism. Gradually this led us to formulate a hypothesis which, while in a sense merely an elaboration of earlier views, differs rather drastically from what is at present accepted. This hypothesis not only seemed to us to promise a better understanding of the causes of at least some forms of autism, but also to offer new prospects for treatment. When we circulated a pre-publication draft of our article among a number of specialists it became clear that, although we received many negative responses, some workers (notably Clancy and McBride, **6**; and Richer, pers. comm.) were thinking along lines very similar to our own. Since there were, moreover, indications that treatments that were in accordance with their and our ideas showed definite promise but were applied in only very few places, we decided to publish our observations and deductions in the hope that they would stimulate further research.

Child–stranger Encounters

Both of us have in the course of over thirty years observed and analysed various types of social encounters in a number of animal species; at the same time we have observed many children, particularly in social encounters, both between children and adults, and of children amongst themselves. One of us (E. A. T.) has gradually turned, whenever her time allowed, to increasingly detailed and systematic observations of such human encounters. On the basis of our observations both of us have further learned to apply certain procedures in meeting children (mainly, though not exclusively, toddlers) and these experiments have invariably induced, in children that responded to us initially by showing mild but unmistakable components of Kanner's syndrome, a rapid change towards normal, often even intensive social behaviour. Although part of the facts we shall describe are undoubtedly known to many observers, we submit that their relevance to an understanding of what really happens in social encounters has so far been insufficiently appreciated, even by the numerous professional child psychologists, psychopathologists, and also ethologists with whom we have happened to discuss our findings.

E. A. T. has found that three semi-controlled situations in particular are admirably suited to the type of observations required: (1) meeting a mother and child during shopping in supermarkets, when, riding on a shopping cart, a child will unexpectedly come face to face with the stranger-cum-observer; (2) sitting near a mother and child on a bus; and (3) either visiting, or being visited by a family with young children. In all these situations one is to the child either a complete or a comparative stranger; the child usually becomes aware of one, and is forced into a social encounter of some kind (the nature of which one can control to a considerable extent); and one can, if certain precautions are taken, observe its behaviour closely. As is usual with observational work, the observer progressively refines the data and their recording, and we are convinced that, however long we have been improving our perceptual competence in this field, a great deal of further refinement is still possible, and will in fact be required, both for the purpose of quantification and for further experimentation.

When meeting a child in any of these situations, the observer (carefully taking the situation and the child's behaviour into account—in fact *monitoring* both all the time) either smiles or simply looks at the child. Its responses are extremely varied. They can range from, on the one hand, definitely positive (such as, for instance, an uninhibited smile, or even approach, and friendly interaction) to, on the other, gaze aversion, and closing of the eyes, turning of the head and even of the whole body, and moving away, often to snuggle up against the mother; and in addition, for instance, thumb sucking. Between these extremes there are numerous intermediate response patterns, some of them extremely subtle—so subtle in fact that, although one can learn to recognise them at a glance, one has to make a quite intense effort of analysis of one's own perception before one can make explicit what one has actually seen (this is, of course, a common and necessary procedure in all observational studies). The observational task is further made difficult not only by the subtlety and number of details of the child's behaviour but also by the (highly relevant) fact that it changes all the time. We have found it essential to pay close attention to these changes, not only for the purpose of registering them, but also for the even more vital purpose of responding to them quickly and correctly. It is of course equally essential to realise that in this type of study one assumes a dual role: that of observer, and that of being part of the input to which the child reacts; one is, so to speak, observer and interferer (or 'dummy') rolled into one.

Initially the child, starting with no more than eye contact (i.e. looking at the stranger *who looks at the child*) can show various types of friendly, socially positive behaviour: subtle expressions of the eyes and body stance which one learns to recognise as, for instance, interested, or friendly, etc.; very slight curving up of the corners of the mouth (often for no more than a fraction of a second); continued eye contact, etc. As already indicated the signs can be, or become, more pronounced and then include, for instance, a progressively more complete, ultimately a 'beaming' smile, positive approaches, and finally a large variety of obviously relaxed, friendly and playful interactions with the observer—incidentally, extremely rewarding and touching behaviours.

The initial, exploratory glance can on other occasions be followed by a negative response pattern. This too involves a wide range of, in part again very subtle, expressions and behaviours. The mildest form of rejecting behaviour is a certain expression of the eyes, which is very difficult to describe but which (as we know for instance from novels)

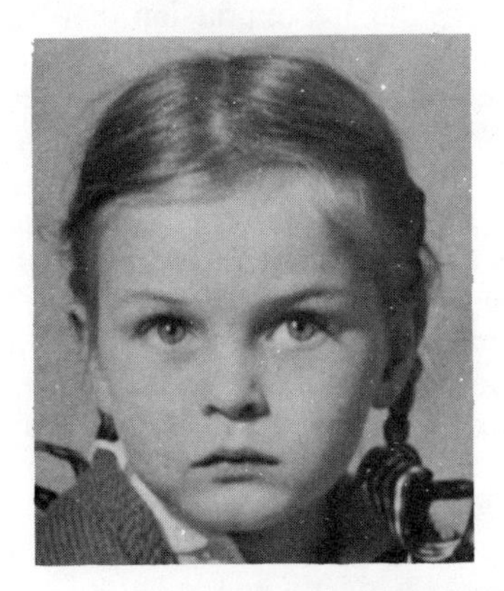

Fig. 155. Above: Two photographs of a girl of 6 taken in the same spring. Left: by a school photographer; right: by her elder sister. Below: Typical friendly eye contact of a 6-weeks old girl held by an aunt unknown to her.

is known to many people. It is often described as an 'empty' or 'blank' expression, also as 'letting the mental shutters down'—it is a vague, expressionless look, often aimed slightly *past* the adult's eyes (an important detail which is easily missed). Anticipating for a moment the issue, to be discussed later, of the powerful communicative effect of the stranger observer's response to this first subtle sign, this initial response of the child is, *if the observer keeps looking at it*, followed by partial or complete closing of the eyes. This can be, but is not always, very slow; sometimes so slow that one has the impression that 'the eye lids have a very long way to go'. When closing the eyes is total, the eye lids look completely smoothed-out—not even slightly creased, pressed, or 'screwed up', as eyes look when closed on other occasions except in sleep.

However, such a child will not keep its eyes closed for long. Sooner or later it will cautiously open them again and look at one. If it finds one still looking at it, the negative response will be repeated. Depending on both the child's condition and the behaviour of the adult, this subsequent behaviour can either be an intensifying of the first response, or a gradual slackening of it, and replacement by the more positive behaviour described above. An increase in the negative response consists of a turning away of the eyes, of the whole head, full or partial closing of the eyes, and moving away, or if the mother is present, the 'snuggling behaviour' already mentioned.

Classification of Observables

It is clear that already in our description we have been classifying a large number of formally very different behaviour elements. This classification involves two procedures.

First of all we have divided the observable behaviours into two major classes, which we named socially positive and socially negative, respectively. Other, roughly equivalent terms are 'bonding' or 'sociable'; and 'rejecting' or 'withdrawing', respectively. The justification for this and similar dichotomies lies in the correlation, within each of the two classes, between (1) recognising as either approaching (positive) or avoiding (negative), and (2) three other, at the time of classifying independent, objective criteria. These are (*a*) the *form* of the behaviour elements (thus the various degrees of closing the eyes or of turning away the head are objectively similar to each other); (*b*) their *sequential association* either with overt approach or with withdrawal (which the trained brain monitors with surprising intuitive accuracy, but which, to avoid bias, must nevertheless be checked ultimately by modern sequential analysis procedures); and (*c*) their *function*: they *result* either in the intensification, or the reducing and even the breaking-off of social contact and proximity. This is because they act as signals to the stranger, which either promote or discourage the establishing of contact *by him or her*. It is, of course, the correspondence between (*a*), (*b*) and (*c*) which makes both behaviours so strikingly and intricately adapted—it ensures their proper functioning in the social context.

Secondly, our classification has included a rating, *within* each of the two classes, of the behaviour elements as ranging from 'mild' or 'strong'—that is, in essence, according to an intensity scale. Thus merely closing the eyes is rated as a less intense negative response than turning away. This classification too, though done intuitively at first, is not arbitrary. It is again based on certain correlations, for instance between the extent, forcefulness, speed, or duration of the behaviour element; the place it occupies in the most frequently observed sequence towards either fully positive or fully negative behaviour; the strength of the stimulus required for their elicitation, etc.

We feel that an awareness of the criteria for such preparatory classifications is to be an essential part of the method if it is to become scientifically acceptable. One often meets with criticism to the extent that these classifications are either 'arbitrary' or 'subjective'. We submit that neither objection is valid—we all use them intuitively in our everyday social interactions, and in fact would socially misfire if we did not. We believe that, as the methods advocated here (and the nature of the proposed correlations themselves) will be more critically examined, they will be found as useful in the analysis of human social behaviour as they have been in that of animal behaviour (see, for instance, the relevant chapters in textbooks such as Manning, **18**; Hinde, **11**).

Complexity and order

While we are for the moment concerned exclusively with the behaviour of the child, it is worth pointing out already here that this forms only a part of an extremely complex, and in fact triangular interaction. We have already said that the behaviour of the stranger is just as important as the momentary state of the child. In addition, the child's mother usually responds, consciously or unconsciously, but often with uncanny perceptiveness, to even slight negative responses of her child, and she will then make a variety of reassuring movements. For instance, a mother will often extend her protective hold on the child, pat or stroke it on its head, smile at it, etc. Also, when the mother is aware of the stranger's conduct, she may either throw her a slightly defensive glance, or alternatively show her a friendly, or an understanding reaction. Finally, the stranger, particularly one with motherly inclinations, finds it surprisingly difficult to keep looking at a negatively responding child; she experiences this (as we shall argue, correctly) as cruelty on her part. This, incidentally, introduces another, technically disruptive element in the already difficult procedure, and one which is hard to eliminate. While the roles of 'stranger' and observer could of course be allocated to two different persons, both of them would always be emotionally involved with the child. This, however, is not only unavoidable in our procedure, it is, of course, at the same time of distinctly positive value because it increases one's sensitivity—in fact we believe that some studies, and even some attempted cures of autistic children suffer from a deplorable lack of sympathy with the child. But it cannot be denied that it is this very sensitivity to signs of distress in the child that does limit the scope of one's studies; it makes one less and less willing to deliberately intensify the child's negative responses. Here, as so often, we have to accept, on moral grounds, limitations of research on human beings. Studies of animals suffer less from these restrictions, and it is therefore relevant to discuss an example of social interaction between animals in which one of *them* is a stranger to the other, and so takes upon itself the (natural) task of being 'cruel'.

First, however, we will give a few examples of the kind of observations on children which have in the course of the years convinced us not only that Hutt and Hutt (**16**) are fully justified in drawing attention to the many similarities between the behaviour of autistic children and that of normal children under certain conditions, but also that a study of the circumstances under which this occurs can be revealing. Of our numerous observations we select three that we consider representative.

1. We once had occasion to observe a 2-year-old child in the surgery of a highly perceptive paediatrician; in fact, one of our own children, which of course carries the risk of being biased, but also the advantage of knowing a great deal about the child's usual behaviour. In the doctor's presence she suddenly showed, much to his distress and (we are ashamed to say) merely to *our* surprise, total withdrawal, alternating with extreme restlessness, a number of stereotypies, and various signs of acute anxiety. Once outside the surgery and on the way home the child soon returned to normal, sociable behaviour. Of course many practising physicians (and certainly many parents) know that this is a very common occurrence.

2. On another occasion this same child, then 5 years old, had to be

hospitalised for two weeks of observation. While in hospital, she was confined to a bed in a children's ward, and was not allowed to move outside the confines of this bed. Whenever we visited her, she showed non-verbal signs of acute distress, and in addition (or rather, as part of this) extreme restlessness, alternating with periods of total withdrawal in a corner; stereotypies as striking as the incessant running in loops of captive wolves, and a complete absence of overt responses to whatever we said.

This child, though always timid, was otherwise 'normal', and has since grown up to a well-adjusted adult, with more than usual manual dexterity (she is a professional string player) and above-average ability in the use of at least three languages, enterprise in travelling abroad on her own, and she is, at 25, a successful teacher even of persons twenty years her senior.

3. Another child (again a close relative whom we have observed from shortly after birth, for periods varying from one to fourteen days) showed already at a very early age a curious mixture of extreme clumsiness and, when interested, equally extreme powers of refined manipulation. Thus, when 5 years of age, he still had difficulty in finding his way into his father's pockets, and on a climbing frame he was easily outstripped by his sister who was 18 months younger; but he could, when not yet 4 years old, put a gramophone needle unfailingly on the groove of a rotating record of a classical symphony *at the place where (as was clear from his humming beforehand) he knew that a certain phrase occurred.* This child had, from the age of a few months, been excessively alert to sounds, and in particular to music; when 5 years old he diagnosed a record, new to him, of a lesser known Haydn symphony correctly as 'Haydn', against the opinion of both (musically sophisticated) parents, who said it was a Mozart symphony. This boy, now 6, is still timid; in slightly unfamiliar situations he often withdraws from social contact; and one of the most striking aspects of this case is that, after his first week at school, the teacher expressed the opinion that he was deaf!

As all students of autism know, this selective cutting-off of part of the available sensory input is a common phenomenon in autists (compare for instance this description, written down when we were still poorly informed about the literature on autism, with Rimland (**21**), pp. 10 and 12). In the literature it is referred to as 'perceptual malfunctioning'; and its erratic occurrence is emphasised. We shall argue that it is in fact not erratic, but concerns special categories of input.

It is observations of this kind, which as we said can be made time and again, that enable us on the one hand to recognise the behavioural deviations, in normals and autists, as parts of withdrawal behaviour; and on the other to correlate their occurrence with particular environmental situations, namely those which are to the child socially and/or physically unfamiliar. It is further striking that such behaviour occurs most in children that are overall shy or timid. It is our experience that a number of child psychiatrists and analysts are reluctant to believe that unfamiliar situations (physical and social) are generally frightening. But a study of 'expressions of emotion' such as we have briefly illustrated leaves no doubt on this point at all. Nor is this surprising to those who are aware of the overall adaptedness of behaviour. In Man as in Primates with a similar (though not identical) affiliation system (such as Rhesus monkeys, see Hinde and Spencer-Booth, **12**) the growing infant begins at a certain age to venture away from the mother and explore the surroundings. It is at this stage that the 'eight months anxiety' of human infants becomes an essential part of adapted behaviour—anything strange *has* to be approached with caution, and the infant *has* to be ready to return to the security of the mother whenever danger threatens—or even simply 'in case of doubt'. The conclusion seems inescapable that Kanner's syndrome is likely to occur temporarily in shy children when they are confronted with, to these children, intimidating situations and that the input they refuse to admit is input that frightens them. If we rely on observed overt behaviour—and this is after all our only immediate source of information—the difference between such children and autists can only be seen as one of degree, and it is no more than straightforward description when we apply to the occasional behaviour of such normal children the term 'temporarily autistic'. We believe that this leads to a better understanding of at least some cases of 'real', more permanent autism.

Motivational Conflicts

Before carrying our analysis further, a few words have to be said about the second field of research in which ethological methods could be of much greater use than at present appreciated: the study, at the non-verbal level, of 'motivational conflicts'. This relatively advanced area of Ethology has produced the significant and well-established conclusion that many movements with a communicating *function* are *caused* by motivational conflicts in the signaller.

Of the large variety of social interactions in animals studied by ethologists those concerning the 'bonding' or 'socialising' are most relevant; in particular the pair bonding in monogamous species, in which male and female establish a personal bond. Of this category, those (numerous) species in which the reproductive season begins by an upsurge of intraspecific aggressiveness between males are the most relevant. Our description, of necessity extremely condensed, will refer to a group of species on which our own work has concentrated: the Herring gull and its relatives, but the overall conclusions are valid for many other species.

In these species the prospective partners go through a number of phrases, of which the first one, that of forming a pair bond long before actual coition takes place, will concern us here; such partners, once bonded, remain for quite a period in a state of 'betrothal'. The tendency to bond is deduced from numerous observed approaches, but it is initially greatly hampered by two other behaviour systems, each advantageous in its own right: attack, and avoidance. The male responds to the approaching female by predominantly aggressive behaviour, due to the fact that, as an intruder of the same species, she presents parts of the stimuli characterising a rival male. Although this aggressive behaviour is mixed with a tendency to approach (and to allow approach by) the female, and to a lesser extent with a tendency to withdraw, it is his hostility that often blocks a female's early approach attempts, for she reacts very sensitively to his expressions of aggressiveness. She shows numerous signs of 'timidity': her attempts to approach the male are mixed with keeping her distance, or even fleeing altogether. It is the female's behaviour that has to concern us here.

Two aspects have to be considered: the continuously fluctuating motivational state of both partners, and the *mutual* signalling *effects* of their movements. We describe the motivational state in terms of the degree of activity of each of the specific, major functional systems. Methods have been developed for identifying these even if the full overt behaviour patterns of these systems are not shown—see, for instance, Tinbergen (**25**) and Manning (**18**). Their application leads to the conclusion that the great variety of behaviours shown by such a female is due to the simultaneous arousal of two behaviour patterns that are wholly or partly incompatible—in this case approach to the male and withdrawal from him.

In some cases this interpretation has been confirmed (and supplemented) experimentally (see, for instance, Blurton Jones (**2**) and Tugendhat (**29**).

While the detailed causation of all the different movements that form part of the observed 'conflict behaviour' is not yet fully understood, it is already abundantly clear that in such cases of 'mixed' or 'ambivalent' motivation it is too simplistic to explain the behaviour exclusively in terms of *general* arousal; what is aroused in many of

Fig. 156. Above: Female **Larus fuscus** (left) 'facing away' in response to a prospective male partner—approach in conflict with withdrawal. Below: Female **Larus fuscus** (right) approaching the male in the early 'betrothal' period; her 'hunched' posture and sideways orientation are typical of a slightly more advanced stage in the bonding process.

these situations is a set of relatively specific behaviour systems. This is not to deny the occurrence of general arousal—that autistic children are generally hyperaroused emerges from the EEG data of Hutt *et al.* (**14**), and has also been suggested by Hermelin and O'Connor (**10**); and the indications are that animals under the influence of ambivalent motivation are also hyperaroused (Delius, **9**). But we submit that thinking in terms of general arousal alone does not go far enough and discards parts of the available evidence. Similarly, it is obviously too simplistic to describe these phenomena merely in terms of overall 'quantity of input'—the quantity of types of input (described in terms of the specific behaviour systems elicited) is probably more directly relevant. This is why we think that the work of Stroh and Buick (**24**), which reports on an already encouragingly successful treatment consisting of 'reducing input', might profitably be refined by explorations of the effects of reduction of more specific inputs, in which for instance frightening stimuli are separated from food, from comforting stimuli, etc.

Returning to the behaviour of gulls: the conflict of the female gull oscillates all the time between more 'daring' approaches and incipient or complete withdrawals; in addition there is a slow, consistent and progressive change leading to a waning of avoidance and a shift towards more predominant approach, ultimately ending in bodily contact, and finally in coition.

Predictions made on the basis of gradually refined procedures have time and again confirmed the general validity of these conclusions, even though a (slowly diminishing) number of details can still take us by surprise. The entire, highly complex overt pattern is illustrated in detail in the film published by Tinbergen and Falkus (**28**)—no amount of verbal description can convey the richness of the phenomena as clearly as this photographic record.

For our present purpose it is of special significance that in such encounters each of the two behaviour systems is in each partner under the control of highly specific external and internal agents, and that the fluctuations in both these systems are expressed by a set of movements of which each acts as a signal to the partner. It is therefore not enough to observe the female only; one has to record changes in the partner's behaviour and in the rest of the environment before one can check the validity of predictions. This procedure of detailed, interpretative observation may lack the glamour of controlled experimentation and of measuring internal, 'physiological' events, yet for the study of a *behavioural* phenomenon the more direct and appropriate method is to systematise the recording of details of the *behaviour*, and of the *situations* that affect it.

Application to children

The application (of these *methods*) to human children, while, as some recent publications show, off to a promising start,[1] is still far behind that already achieved with some of the better known animal species. When we applied them to child–stranger encounters in the three situations already described, we were forced to the conclusion that an encounter between a child and a strange adult creates in the child a state of motivational conflict that is very similar to that described for female gulls in the pair formation process, except that of course the approach tendency is not sexually motivated. The children show by their overt behaviour that they are attracted to the stranger, i.e. tend to make eye contact, to approach him or her, and to further engage in social interaction of a variety of types.[2] At the same time they are definitely afraid, i.e. reluctant to approach. As we have seen, this avoidance tendency expresses itself in a rich syndrome of more or less concurrent behaviour components, such as keeping distance or moving away, avoiding eye contact, various degrees of closing the eyes, slightly or entirely turning away the head or the whole body, etc., etc. The sensory modality that is 'locked out' may be different from one child to another; most children seem to avoid primarily visual input, but many reject auditory and some even tactile input, but it is always *intimidating* input. We have to stress once more that for our understanding of the child's state it is of paramount importance to realise that the relevant input is neither general, nor modality-specific, but *behaviour*-specific; and the inputs that count seem to be those that specifically stimulate or reduce either withdrawal, or approach, or both. A child that seems 'deaf' may, as we know from detailed experiments, be at the same time extremely sensitive to, and attracted by, non-social noises such as the faint ticking of a clock, the calling of a bird outside, etc. A child that averts its gaze from us may at the same time be highly alert to the visual properties of its toys. This means that those who apply the general terms 'perceptual malfunctioning', 'quantity of input', and 'general arousal' are in effect not only discarding, but *even contradicting* a great deal of highly relevant evidence.

In short, we submit that the children described above, and according to the available evidence many autistic children, are all willing, indeed *very keen* to establish a social bond, but are prevented from

[1] See, for instance, the volume edited by N. G. Blurton Jones (1972).

[2] Positive seeking of social contact becomes for instance excessive when, in playing 'hide and seek' with a child, one deliberately fails to find it for too long; this elicits frantic appeals for contact.

showing the appropriate behaviour by inhibition through fear. In normal children, even those that are 'temporarily autistic', the initial conflict is quickly solved (as in the female gulls) if the stranger does a combination of two things: (1) avoids stimulating withdrawal; and (2) positively enhances approach and contact. The extremely variable patterns of behaviour shown by such a child are due to moment-to-moment fluctuations in its state (requiring continuous adjustments by the adult); these quick fluctuations are in part due to the external situation and may in part reflect quick internal switches, although it is often difficult to be sure that the environment has remained constant. Most adults, including many professionals, fail to see that very slight variations in the behaviour of the intimidating partner can influence the state of the recipient, and can tip the balance between withdrawal and approach. A straight look at a timid child, *even when one is smiling*, may in a temporarily or permanently 'cowed' child elicit increased timidity, whereas in a less anxious child the objectively identical signal enhances a positive response. All this is a simple matter of fluctuating thresholds, but the social process appears so exceedingly variable because the standard by which the child responds is continually fluctuating. And it is the child, not the 'objective' external situation which is the relevant standard. It is often failure of the observer to 'gauge' this fluctuating state of the child which makes the behaviour of the children so puzzling to him.

Experimental Testing

It is, of course, inherent to the nature of this work that the conclusions suggested so far are no more than interpretations of what amounts to clinical evidence. We shall now describe experimental evidence in the form of a procedure which we have been applying for a long time when meeting an initially distrustful child. *Our application of this procedure has invariably led to a remarkably quick establishment of a strong and long-lasting social bond.* In order to appreciate the nature of the experiment one has to keep in mind that looking at a child (even when one is smiling, and certainly when one makes more vigorous, friendly motivated movements such as approaching, exclaiming, and generally 'crashing through the child's barrier') has an intimidating effect as long as the child is still socially negative. One has to adjust one's signalling behaviour from moment to moment to the signs of conflict in the child, and do so in such a way that one just avoids increasing the withdrawal tendency, and will be *interpreted* by the child as both non-intimidating and positively friendly. As remarked previously, these actions are not simply arrangeable on one scale; one has to do specific things to prevent the child from withdrawing, and different, but equally specific things that make it seek contact. One has also to remember that social contact involves much more than mere visual stimulation; that in fact with most negatively reacting children certain forms of tactile and vocal communication are far less intimidating and more inviting than eye contact, which remains frightening up to an advanced stage in the bonding process. And the visual stimuli themselves involve of course much more than eye contact. In dealing with a toddler it is for instance important to squat, so as to bring one's eyes level with those of the child. Other positively bonding procedures are: joining the child in something that interests it at the moment (and communicating this non-verbally); if possible laughing *with it*—at any age laughing together *at* something or someone is one of the most strongly bonding behaviours known.

What we invariably do when visiting, or being visited by a family with young children is, after a very brief friendly glance, ignore the child(ren) completely, at the same time eliciting, during our initial conversations, friendly responses from the parent(s). One can monitor a surprising amount of the child's behaviour by observing out of the corner of one's eye.

Usually the child will start by looking intently at one, studying one guardedly. One may already at this stage judge it safe to now and then look briefly at the child to assess its state more accurately. If the child averts its glance, one must at once look away. Very soon the child will approach gingerly, and it will soon reveal its strong bonding tendency by touching one—usually by putting its hand tentatively on one's knee. This is a crucial moment: one must *not* respond by looking at the child (which may set it back considerably) but by cautiously touching the child's hand with one's own. Playing this 'game' by if necessary stopping, or going one step back in the process (adjusting to the child's response) one can soon give a mildly reassuring signal *by touch*, for instance by gently pressing its hand, or by touching it quickly and withdrawing again. If, as is often the case, the child laughs at this, one should laugh oneself, but still without looking at the child. Soon it will become more daring, and the continuation of contact-by-touch and by indirect vocalisation will begin to cement a bond. One can then switch to the first, tentative eye contact. This again must be done with caution, and step by step; certainly with a smile, and at first for brief moments. We find that first covering one's face with one's hands, then turning towards the child (perhaps saying 'Where's Andrew?' or whatever the child's name) and then very briefly showing one's eyes and covering them up at once, is very likely to elicit a smile, or even a laugh. For this, incidentally, the child often takes the initiative (see e.g. Stroh and Buick, **24**). Very soon the child will then begin to elicit this; it will rapidly tolerate increasingly long periods of direct eye contact, and join one. If this is played further, with continuous awareness of, and adjustments to, slight reverses to a more negative attitude, one will soon find the child literally clamouring for intense play contact—a virtual rebound of social behaviour. Throughout this process (which is in reality a long *series* of refined experiments) the vast variety of expressions of the child must be *understood* in order to monitor it correctly, and one must oneself *apply* an equally large repertoire in order to give, at any moment, the best signal. The 'bag of tricks' one has to have at one's disposal must be used to the full, and the 'trick' selected must whenever possible be adjusted to the child's individual tastes.

Once established, the bond can be maintained by surprisingly slight signals; a child coming to show proudly a drawing it has made is often completely happy with just a 'How nice, dear' and will then return to its own play.[1] Even simpler vocal contacts can work; many children develop an individual (seemingly nonsensical) contact call, to which one merely has to answer in the same language.

The results of these experiments played in continuous adjustment to the monitoring results, have invariably been surprisingly rapid, and also consistent. Of course, different children require different starting levels, and different tempos of stepping-up. It is also important to the child how familiar the physical environment is in which it meets the stranger. And many children take more than one day; with them it is important to remember that one has to start at a lower level in the morning than where one left off the previous evening. We have the impression that the process is on the whole completed sooner if one continually holds back until one senses the child longing for a more intense contact.

We have found that the procedure here described works very well with many dogs too. In spite of the different forms of social behaviour of dogs the course of interaction is basically the same: by ignoring a dog while engaged in a friendly chat with its owner one soon causes it to put a forepaw on one's knee, and by avoiding too early eye contact while giving the right touch and vocal stimuli one achieves

[1] We think that Bowlby (**4**), in his important book on attachment, underrates the bond-maintaining effects of such fleeting and distant contacts; this is also pointed out by Rheingold and Eckerman (**20**). Bonding can even be maintained 'by proxy'—for instance by working mothers leaving a snack for a child that returns from school before she herself comes home from work.

(even without providing the doubtlessly equally important chemical stimuli) the same 'clamouring for attention' as shown by children, including the presenting of toys.

While there is no space to substantiate our claim that our interpretation applies not only to toddlers but to all age classes, we should like to point out that neither timidity nor bonding behaviour are absent in babies of the pre-crawling age, although their repertoire, and selective responsiveness to stimuli, are different. To a baby in arms or lying in the cradle one communicates friendliness by smiling at it, and speaking in a cooing voice. Contrary to what some mothers have reported, baby anxiety does not in our experience express itself by gaze aversion 'from birth on'; at least it could not possibly be diagnosed so early. But very young babies when frightened for instance by a fall, or by hearing distress shrieks, do stiffen up, arch their bodies, cry, etc. We have seen nurses 'play' with newborn babies by dropping-and-catching them ('when they spread their arms so funnily'); this may well on occasion be a traumatising experience. It even seems to us not at all improbable that in a 'battery parturition' ward a baby *in utero* responds unfavourably to the uncontrolled screaming of some mothers. Evidence of effects of prenatal experience is accumulating all the time (see Carmichael, **5**).

As we have already indicated, the expression of anxiety and the bonding behaviour both change dramatically with the onset of the crawling stage, when the 'eight months' anxiety' appears. It is not superfluous to point out that the behaviour of the mother undergoes corresponding changes; thus her response to distress signs becomes richer, and more suited to the needs of a crawler; when playing with such a more mobile child she will not only squat near it, but also turn herself more or less parallel to it, and look at its hands or toys rather than at its face, etc. etc. The patterns of reciprocal bonding behaviour have undoubtedly a highly adapted 'deeper structure', of which the details will deserve further study. In this context much can be learnt from work on mother–infant interactions in other animals, above all Primates (see, e.g. Crook, **8**).

Our evidence also suggests that in order to understand even the 'temporarily autistic' children, and the history of pathologically autistic children as well, it is not sufficient to study merely the behaviour of these children themselves. While of course some children may genetically be more liable to intimidation than others (see for instance Keeler, in Rimland, **21**; and Rutter, **22**), and the bonding tendency may also vary genetically, the statement (in the report of the Eisenberg Committee, see Wing, **30**) that autism has nothing to do with the behaviour of the parents is not only premature; it contradicts many facts. But we certainly need more information of the required type about parent–infant interaction in Man.

In assessing to what extent parental behaviour might contribute to the autistic deviation, insufficient attention is also paid to the fact, in our view of great importance, that many mothers unwittingly respond to the modern stressful living conditions by behaving either over-intrusively or sub-motherly towards their own children, or by allowing them to be overexposed to unfamiliar situations, and/or by eliciting, and even soliciting over-intrusive behaviour in visitors. We believe that a systematic comparison of parental, and in particular maternal behaviour, of modern Western urban populations and populations of more relaxed cultures might well produce some revealing results.

Infant–infant encounters

One instance of an encounter between normal children must be related here to illustrate both motivational and signal-function aspects of gazing-at and being-gazed-at. E. A. T. found herself sitting in a bus opposite two mothers sitting side by side. One mother had two children, aged at a guess approximately 3 and 4 years old; the other had one, slightly older girl on her lap. Both mothers were staring over their children's heads and did not take any notice of the observer. The siblings of the first mother were enjoying a game: they repeatedly bent over towards each other until they tickled each other with the furry linings of their anorak hoods. After a tickling-and-giggling bout they would slowly draw apart and look at each other with quiet but intense delight, and after a while resume the game. The girl on the second mother's lap was deeply interested. She leaned forward towards the other children as far as she could, slightly turning her head in an effort to observe the others. When, in one of their separating bouts, the siblings happened to notice this, they drew themselves up a little and then looked fixedly at the stranger. This child responded first by smiling at them, but, when encountering two pairs of staring eyes, withdrew gradually till she was leaning backwards against her mother. While withdrawing she stopped smiling. As she withdrew, the siblings leaned over towards her as far as they could. They then ignored the stranger, and resumed their game. However, the strange child could not resist looking again. This time the siblings, on drawing apart, looked immediately in her direction, and now the approach of the two 'insiders' and her withdrawal were quicker than the first time. This process was repeated some six times, after which the outsider stopped looking at the siblings altogether. The observer had the impression that she was simply cowed. She then happened to look at the observer, and reacted to the latter's quiet smile with a reluctant and shy smile of her own. She then alternated between half-smiling at the observer, and looking in a furtive, but keenly interested way at the siblings. We emphasise that this girl always withdrew *before* the siblings actually began to lean over towards her—she was clearly intimidated by seeing the two pairs of staring hostile eyes.

Of course this type of interaction is not at all confined to children, nor to our species; it is very common in various social Primates, where it has even been ritualised: two top-ranking Baboons for instance (whether both males or male and female) can perform the dual function of bonding and threatening-off another individual by alternating looking-at-each-other with together-looking-fixedly-at-the-outsider; this is done with jerky movements accompanied by changes in facial expression. In humans this may well be cross-culturally constant.

Causation, Prevention and Curing

It will be clear by now that we believe not only that autistic children as well as temporarily cowed normal children suffer from a conflict between sociality and fear, but also that the environment has more to do with both temporary and permanent aberrations than is generally assumed. It is quite possible that, as for instance Rutter (**22**, **23**) has suggested, a variety of disturbances are at the moment lumped together under the term autism, and that either genetic, or somatic brain damage does account for a number of cases. But the direct evidence for this is extremely meagre. What cannot be doubted is that genetically or phenotypically timid children will be more at risk than others, but a large proportion of autists may well be children who would have escaped unharmed in less stressful societies. Further, the sharp distinction between permanently and occasionally autistic children does not necessarily contradict our suggestion of a gliding scale—if our ideas about the importance of social environment are even partly correct, there should be very few children who keep hovering between 'normal' and autistic, for both positive socialisation and withdrawal must then be assumed to be self-reinforcing processes, and a discontinuity is bound to develop in what may at the start have been a continuous, graded scale.

From our suggestions about the causation of autism certain ideas about prevention follow directly, and they are clear from what we have said so far: we believe that an understanding of non-verbal expressions of withdrawal, and, based on that, carefully adjusted parental behaviour that aims at reducing anxiety and promoting socialisation will be important. The step from here to the recommendation of particular forms of therapy is of course much more speculative. Firstly, autistic children seem often to develop into virtual 'tyrants' of the family by the inevitable attempts of all concerned to prevent at all costs outbursts of 'temper tantrums' (which we believe to develop from what initially are really 'panic tantrums'). A re-establishment of some kind of discipline may therefore be a first requirement. Secondly, a cure may not necessarily consist of reversing the course of the affliction's history. A third difficulty has to do with the uncertainty about the question which parts of the syndrome are primary, and which aspects secondary. It will be clear from our account that we believe that for instance speech defects, stereotypies, and general arousal are secondary developments or 'symptoms' due to the unsolved motivational conflict, and that therefore speech therapy (which often involves behaviour experienced by the child as frightening); enforced teaching of social skills beyond the patient's mental age; and certainly electro-convulsive shock therapy are not likely to be very effective, and may actually cause an *overall* deterioration. And in fact such treatments seem so far to have led to little success. What we have seen and heard so far suggests that patient restoration of a sense of security in the child (perhaps preceded by quiet, impersonal re-establishment of discipline), combined with an involvement of the mother in the cure (as done with apparently very good results by H. Clancy, see Clancy and McBride, 6) is the most promising course to take. And this course would logically follow from our hypothesis about causation. We ourselves have hardly any experience, but it is worth mentioning that we have not only observed some successful professionals in action and found that they did exactly what our hypothesis would require, but have also on a few occasions tested out responses of autists to our own behaviour, either when we behaved in a mildly intimidating way or, when we 'socialised' according to our 'taming' procedure as described above.

In one particularly striking case an autistic boy, who had been sitting on a nurse's lap, made his way through a room with some twelve mentally disturbed children (of whom two others had also been diagnosed as autistic) and backed onto N. T.'s lap, soon joined him in mutual hand-touching games, and stopped the incessant teeth-gnashing which had literally hurt our ears while he was on the nurse's lap. He was used to her; she was well-intentioned *but could not stop herself from smiling at him in a motherly, loving way* (as the nurse in Hutt and Hutt, **16**, Figs. 9–12).

It has only gradually dawned upon us that many persons dealing with children, and even professional child psychologists and psychiatrists, are not aware of the intimidating effects of too intrusive approaches. It may be because of this that our 'taming' procedure, so simple (and so often applied intuitively by non-professionals) has hardly ever been systematically tried out.

Conclusion

In this paper we have been acting on the conviction, which is being expressed with increasing insistence by many ethologists, that our understanding of human behaviour can be promoted by the application of Man of certain methods that have been developed in animal ethology. Following in the footsteps of that great pioneer of child psychology, Maria Montessori, and, with respect to our particular subject, continuing in the direction recommended by Hutt and Hutt (**16**) we have considered what observations and simple experiments on the causation of non-verbal behaviour of normal and autistic children could tell us about the determinants of the autistic process. We found strong indications that many autistic children suffer from a form of stress caused by an unsolved conflict between over-timidity or fear of the unfamiliar on the one hand, and frustrated sociality on the other, and that such stress might well, in predisposed children, be largely the result of environmental factors, among which parental behaviour is of primary importance. We consider as insufficiently supported the claim that autism is not due to parental influences but must as a rule be exclusively due to either genetic or organic brain defects. We should like to add that to say that parental behaviour may *contribute* to the development of autism is not to *blame* parents; we have seen many cases of what we consider deviant parental behaviour in couples, particularly mothers, who live themselves under stress, for instance in attempting to combine an intellectually demanding professional career with motherhood.[1] If we read the signs correctly the recent increase in incidence of Early Childhood Autism—which we believe to be largely real, rather than due to improved diagnosis—may well be a consequence of increased social stress.

As far as can be judged at the moment, the forms of therapy that

[1] There are indications that a warm, cheerful family atmosphere, in which parents share the children's sense of humour and light-hearted fun is of great importance.

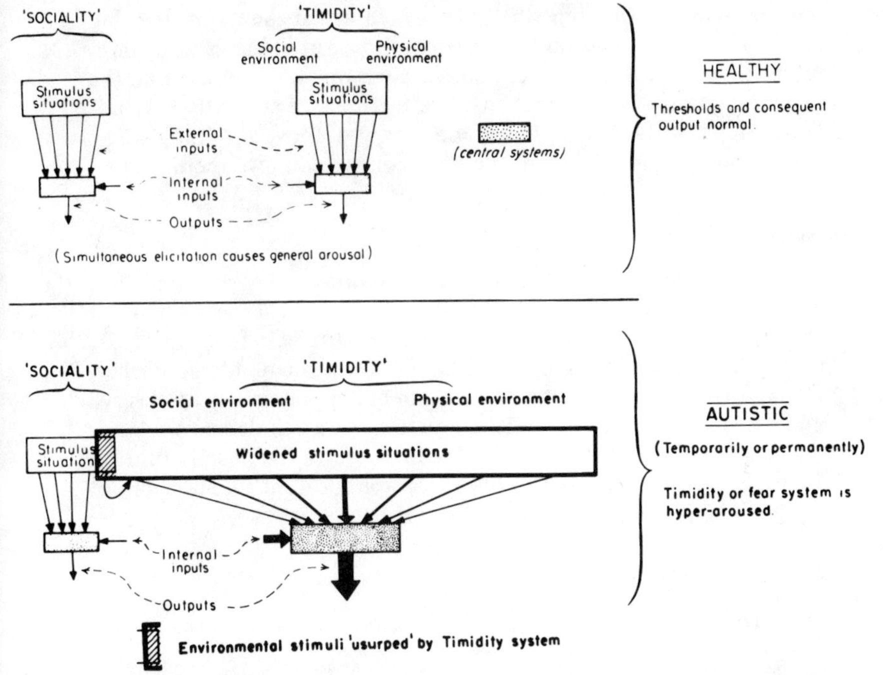

Fig. 157. A greatly simplified one-way 'flow diagram' of the motivational aspects of our hypothesis. Feedbacks, output copies, interrelations between sociality and timidity (and of both with, e.g. exploration, aggression, feeding, etc.) left out: details of stimulus situations, of internal inputs and of outputs not given; secondary effects (loss or regression of speech; stereotypies) discarded. The extension of 'effective stimulus situations' (= lowering of respective thresholds) fluctuates with the level of behaviour-specific arousal as well as with overall conflict-arousal—temporarily unbalanced in many normal children; more permanently and progressively fear-oriented in autistic children. Whether or not the sociality system remains constant or shows higher arousal as the result of its frustration has been left open, though its 'rebound' when social contact has been established suggests the second view. Attention is called to the cross-hatched area in the lower diagram: the input into the over-sensitive timidity system is partly 'stolen' from what would normally be input into the sociality system; hence even a friendly face can elicit a fear response in the over-timid child.

lead to partial or total recovery seem to be well in accordance with these views about the causation of autism.

We are well aware of the tentative nature of our hypothesis and of our views on the different therapies which are at the moment being tried out, but our decision to publish our ideas at this stage is prompted by the considerations that many experts on autism admit that the phenomenon is still baffling; that the methods of study that we advocate have so far hardly been applied; and above all that autistic children desperately need all the attention we can give them.

Acknowledgements

We owe a great deal, for stimulating discussions and correspondence, to the following colleagues: Dr N. G. Blurton Jones, Dr John Bowlby, Dr F. Hall, Dr C. Hutt, Dr S. J. Hutt, Dr L. Kanner, Dr G. J. van Lookeren Campagne, Dr G. O'Gorman, Mr J. Richer, Dr G. Stroh and Dr L. Wing. For valuable opportunities to observe and discuss a number of autistic children we are indebted to Mr J. Richer, Mrs B. Furneaux and Mrs B. Roberts. Dr N. G. Blurton Jones kindly allowed us to consult pre-publication manuscripts of chapters of *Ethological Studies of Child Behaviour*.

REFERENCES

1 BETTELHEIM, B. (1967). *The Empty Fortress: Infantile Autism and the Birth of Self.* London, Collier-Macmillan.

2 BLURTON JONES, N. G. (1959). 'Experiments on the causation of the threat postures of Canada geese', 11th Rep. Wildfowl Trust, 46–52.

3 —— (Ed.) (1972). *Ethological Studies of Child Behaviour*. London, Cambridge University Press.

4 BOWLBY, J. (1969). *Attachment and Loss.* Vol. 1, *Attachment.* London, Hogarth Press.

5 CARMICHAEL, L. (1970). In *Carmichael's Manual of Child Psychology*, ed. P. Mussen, Chapter 6, 447–563.

6 CLANCY, H. and G. MCBRIDE (1969). 'The autistic process and its treatment', *J. Child Psychol. Psychiat.*, **10,** 233–44.

7 CREAK, M. (Chairman) (1961). 'The schizophrenic syndrome in childhood. Progress Report of a Working Party', *Br. med. J.*, **2,** 889–90.

8 CROOK, J. H. (1970). 'The socio-ecology of primates', in *Social Behaviour in Birds and Mammals.* J. H. Crook (Ed.) London–New York, Academic Press. 103–59.

9 DELIUS, J. D. (1967). 'Displacement activities and arousal', *Nature, Lond.*, **214,** 1259–60.

10 HERMELIN, B. and N. O'CONNOR (1968). 'Measures of the occipital alpha rhythm, in normal, subnormal and autistic children', *Br. J. Psychiatry*, **114,** 603–14.

11 HINDE, R. A. (1970). *Animal Behaviour*. 2nd edn. New York, McGraw-Hill.

12 —— and Y. SPENCER-BOOTH (1971). 'Effects of brief separation from mother on Rhesus monkeys', *Science, N.Y.*, **173,** 111–18.

13 HUTT, C. and S. J. HUTT (1968). 'Stereotypy, arousal and autism', *Psychol. Forschung*, **33,** 1–8.

14 ——, S. J. HUTT, D. LEE and C. OUNSTED (1964). 'Arousal and childhood autism', *Nature, Lond.*, **204,** 908–9.

15 —— and C. OUNSTED (1970). 'Gaze aversion and its significance in childhood', in *Behaviour Studies in Psychiatry*. ed. S. J. Hutt and C. Hutt, Oxford, Pergamon Press, 103–21.

16 HUTT, S. J. and C. HUTT (1970). *Direct Observation and Measurement of Behaviour*. Springfield Ill., Charles C. Thomas.

17 KANNER, L. (1943). 'Autistic disturbances of affective contact', *Nerv. Child*, **2,** 217–50.

18 MANNING, A. (1972). *An Introduction to Animal Behaviour*. 2nd edn. London, Edward Arnold.

19 O'GORMAN, G. (1970). *The Nature of Childhood Autism*. London, Butterworth.

20 RHEINGOLD, H. L. and C. O. ECKERMAN (1970). 'The infant separates himself from his mother', *Science, N.Y.*, **168,** 78–83.

21 RIMLAND, B. (1965). *Infantile Autism*. London, Methuen.

22 RUTTER, M. (1968). 'Concepts of autism: a review of research', *J. Child Psychol.*, **9,** 1–25.

23 —— *et al.* (1971). 'Autism—a central disorder of cognition and language?' Prepublication copy of a paper given to the Study Group on Infantile Autism. London.

24 STROH, G. and D. BUICK (1970). 'The effect of relative sensory isolation on the behaviour of two autistic children', in *Behaviour Studies in Psychiatry*. ed. S. J. Hutt and C. Hutt, Oxford, Pergamon Press, 161–74.

25 TINBERGEN, N. (1959). 'Comparative studies of the behaviour of gulls (Laridae); a progress report', *Behaviour*, **15,** 1–70.

26 —— (1964). 'Aggression and fear in the normal behaviour of some animals', in *The Pathology and Treatment of Sexual Deviation*, ed. I. Rosen. Oxford, University Press, 2–23.

27 ——, E. ENNION and H. FALKUS (1970). *Signals for Survival*. Oxford, Clarendon Press.

28 —— and H. FALKUS (1970). *Signals for Survival*. Documentary film. New York, McGraw-Hill.

29 TUGENDHAT, B. (1960). 'The disturbed feeding behaviour of the Three-spined stickleback', *Behaviour*, **16,** 159–87.

30 WING, L. (1970). 'The syndrome of early childhood autism', *Br. J. Hosp. Med.*, Sept. 1970, 381–92.

(*The Croonian Lecture, 1972*)

18
Functional Ethology and the Human Sciences 1972[1]

As long as a man feels healthy and happy he tends to take his condition for granted. It does not occur to him that living is like a tightrope act—that there are infinitely more ways in which one can fail than the one narrow road that leads to success.

Yet when he happens to be a field biologist who observes animals in their natural habitat, he realises only too clearly that their life is so to speak a multi-dimensional tightrope act; he sees all the time that their success depends on coping with a bewildering variety of obstacles in the environment, of environmental pressures.

We all have grown up in times of rapidly increasing efficiency, of growing mastery over our environment, and so of increasing security. But now we begin to see that this security cannot be taken for granted —that we have been lulled into an attitude of complacency, and even of arrogance—and we are feeling an uneasy sense of doubt. Reluctantly we are waking up to the fact that we are still, or rather again, threatened by our environment, and a growing number of 'prophets of doom' believe that we are in serious danger of floundering (Ehrlich, **25**; Commoner, **9**, **10**; Goldsmith *et al.*, **31**; Ward and Dubos, **92**). This feeling of unease is the more disconcerting because, unlike a tightrope walker, we do not know the nature of the threats.

It is therefore not astonishing that many biologists begin to feel the need for a closer study, not just of living mechanisms as such, but of the relationship of whole animals, the carriers of these mechanisms, with their environment—that particular relationship we call adaptedness. I am one of those biologists—I have always taken delight in observing the behaviour of animals in their natural environment (Tinbergen, **82**, **86**). And although I have that feeling, shared by many scientists, that the more I see the more I am aware of

[1] Delivered on 18 May 1972.

how much I do *not* understand, I want to try in this paper to sketch how an animal ethologist studies adaptedness of behaviour; how this makes him look at some problems facing modern Man; and how he wonders about the potential value of his approach to these human problems. For this I have to cover a great deal of ground, and my treatment of details will have to be sketchy.

Genetic and Cultural Evolution

It has often been stressed (e.g. by Huxley, **41**; Medawar, **61**; Dubos, **24**), but it bears repetition, that any comparison of the behaviour of Man with that of animals is off to a false start unless the distinction between genetic evolution and cultural evolution is clearly seen. Not until this distinction is made can we fruitfully consider how one relates to the other.

The difference is not to be found in the phenotypic transfer itself of knowledge from one generation to the next. It is true that we rely more on such socially induced behaviour programming than any other species, but such 'cultural' transfer does occur in many animals. What *is* peculiar to our species is that each generation adds to this supplementary programming—that the transfer is cumulative, and accelerating. It is this accumulation that leads to evolution, and for cultural *evolution* there is no precedent—it is a totally new experiment of nature, and we are the guinea-pigs.

Genetic and cultural evolution have in common that they are changes towards adaptedness, or maximum success—success in the difficult task of coping with the specific requirements imposed by the 'niche' that each species occupies. As every ecologist knows, these niches, and their particular pressures, differ from one species to another.

In its early stages, cultural evolution led to great strides in the process of adaptation. It enabled Man to improve his hunting techniques, then to develop agriculture; to protect himself from even the largest predators; even to cure illness. But apart from these short-term advantages, the cultural evolution is also having long-term consequences. It is some of these consequences that now begin to cause alarm. Unlike the numbers of animal populations, which oscillate, those of our own species have shown consistent growth for a very long period—the term 'population explosion' is a mere description of the steep acceleration of this process in recent times. Unlike animal populations, we use up the capital of a number of finite natural resources rather than their interest. Finally, numerous kinds of waste that we produce are not ploughed back and recycled.

In the process we are doing something done by no other species: instead of merely maintaining ourselves in our niche, we are literally conquering our environment; we are changing, on a global scale, the very habitat in which we live. We are creating a habitat that diverges more and more and with increasing speed from that to which genetic evolution has adapted us.

Changes in the environment are nothing new. All organisms have been subjected to them, either when overall climatic changes modified their habitat, or when they themselves invaded new habitats. They have responded to such changes either by genetic re-adaptation, or by moving to new areas of a suitable habitat, or they have failed and become extinct.

The fundamentally novel effect of human evolution is that the anthropogenic change of our habitat is now proceeding so fast that genetic evolution cannot possibly keep pace with it—each of us becomes a Rip van Winkle even in his own lifetime (Fig. 158). We have coped with this in three ways: by invading more and more land areas of the world; by living in far higher densities than our ancestors; and by adjusting phenotypically to a wide range of new conditions.

Many people think that this is a splendid thing, and they equate the rapid change with 'progress'—a loaded term that assumes that we are becoming steadily more successful. And it must be admitted that, judged by the crude biological standard of numbers of our population, Man has so far been extremely successful. But it is worth asking whether we are not overshooting the mark.

Our expansion over inhabitable new areas is coming up against the limits of 'Spaceship Earth' and it becomes clear that our only option is to rely more and more on phenotypic adjustability. What an increasing number of 'prophets of doom' are urging us to consider is whether this adjustability, which is undeniably very much greater than that of any other species, is not in danger of being overstretched. Can the cultural evolution continue unbridled, or will our survival, or at least our wellbeing be threatened?

The question can no longer be ignored because, beyond the short-term benefits of medical, food-producing and physical technologies the long-term effects of our conquest are clearly beginning to bite. One may doubt whether or not all the details of the barrage of warnings and alarums to which the 'prophets of doom' are subjecting us are correct, but their general thesis cannot possibly be shrugged off: there is no denying that a large proportion of the world's population has not enough to eat; nor that car exhausts are damaging our health; nor that certain nitrogenous fertilisers are doing harm to our soils

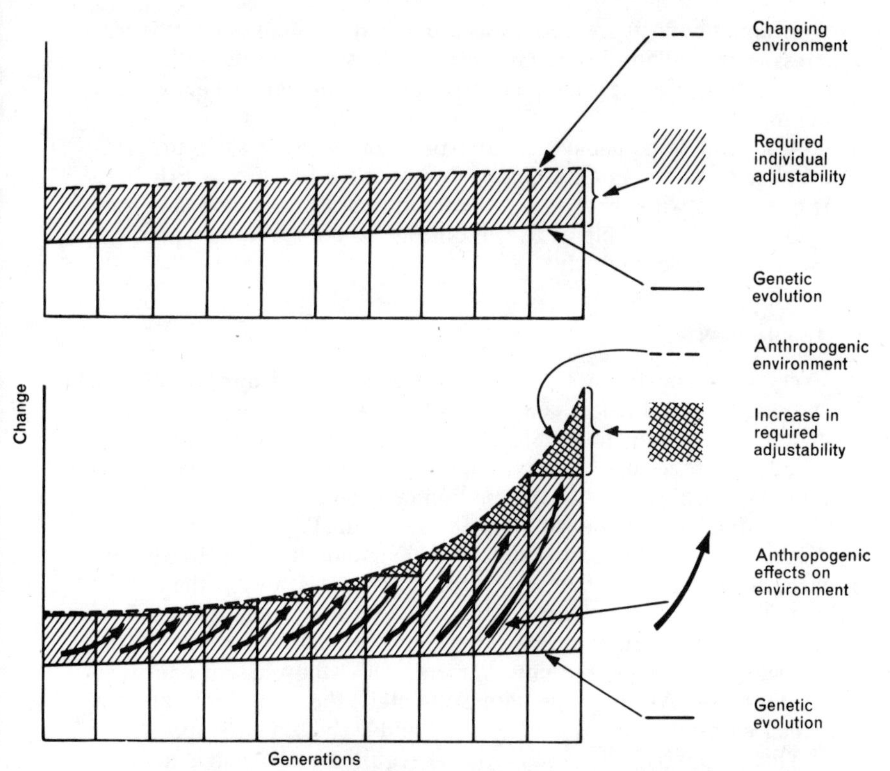

Fig. 158. Schematic representation of the difference between genetic (above) and cultural evolution (below), to illustrate the rapidly increasing demands imposed upon individual adjustability in Man. Further explanation in the text.

and inland waters and our food supplies; nor that numerous industries dump harmful waste on a rapidly increasing scale—the list of deleterious influences is growing almost daily. This material backlash of our culture, and its harmful somatic effects are so striking that they make front-page news. But serious though they are, these somatic effects will not concern me here; I will confine myself to scrutinising the much less tangible behavioural consequences.

The student of behaviour is involved in this issue for more than one reason. First, the cultural evolution is a behavioural evolution—it is our behaviour that has changed our environment, and with it the relationship between what we are doing and what the new environment requires from us. Secondly, although the material changes and their somatic effects are seen most easily, there are signs of behavioural stress as well. Thirdly, the prevention of possible disadaptation and the creation of a new adaptedness will be a matter of behavioural planning.

Adaptedness of behaviour, and programming for such adaptedness are tasks facing all animals. Comparative Functional Ethology is concerned with these very issues, and a brief excursion into this field might at least help us to see some aspects of the human predicament a little more clearly.

Adaptedness

Every time one follows the life of a generation from its start to successful reproduction, and sees how few individuals achieve this end, and how many fall by the wayside, one marvels at the intricacies of adaptedness—of the ways in which animals surmount the innumerable obstacles that the environment puts in their way. It is a fascinating occupation to analyse the animal's achievements—to discover in ever-increasing detail the demands imposed by the environment and to confront these requirements with what the animal actually does to meet them. For this, one has to study the effects rather than the causes of behaviour, and to compare them with the pressures exerted by the environment. This study has of course to relate the behaviour to the natural habitat—the situation in and for which natural selection has moulded and stabilised each species.

This comparison between what is required and what is actually done can start at either end of the cause-effect chain: one can single out an isolated behavioural characteristic and ask 'what is this for?', 'what does it achieve?', or one can concentrate on an environmental pressure and ask how the animal meets it. The evidence that illustrates the procedures is widely scattered in the literature, and I confine myself to a few representative samples.

SURVIVAL VALUE OF SINGLE BEHAVIOURS

When taking the first course, one has to see if and how success would be influenced (positively or negatively) if the animal behaved differently from the way it does; if it deviated from what natural selection has designed it to do.

Of course the function of a behaviour pattern can often be seen at a glance. When the fish *Balistes* overcomes the spiny defence of the sea-urchin *Diadema* by aiming a jet of water at its base so that the urchin is turned over, thereby exposing its vulnerable ventral side

(Fricke, **29**), the overall function of this remarkable behaviour is at once clear. Even so, detailed study of such an 'adaptation' will, as always, reveal a variety of refinements.

Similarly, while the overall function of incubation in birds is equally clear, it took years of intensive study of one single species to understand the effects of all the details of the system, their timing and orientation, and their numerous subtle regulatory devices (Baerends and Drent, **2**).

The effects of other behaviours are so difficult to trace that even now no more than inspired guesses are possible; for instance, why do aquatic bugs and beetles carefully wipe off the surface of their eyes before flying off; what is the function of the 'anting' of so many birds (Simmons, **74**); what is the purpose of the curious 'sky-pointing' of a gannet (Nelson, **63**)?

And guesses can be wildly wrong: for a long time the 'trumpeting' of honey bees was thought to be an acoustical signal; now it is known that its main functions are ventilation, evaporation, cooling and chemical signalling (Ribbands, **68**).

Hypotheses about the function of aspects of behaviour usually come from a systematic comparison of as many different species as possible, and a search for correlations between certain environmental features and peculiarities of behaviour. This method has been applied for many years by bird ecologists and ethologists (Lack, **50**; Tinbergen, **83**, **85**; Crook, **13**, **14**); numerous remarkable instances have been reported in a variety of animal types by Lorenz and his co-workers (Eibl-Eibesfeldt, **27**); and of particular interest is the rapid progress in the comparative functional study of Primate behaviour (Crook, **15**).

Experimental work in this field is hampered by the enormous amount of uncontrolled, and often uncontrollable 'noise' found in the natural environment. But one always tends to forget that working under more controlled conditions in the laboratory, so necessary for the analysis of the living machinery, hampers the study of behaviour effects, and makes it impossible to relate one's findings to reproductive success in nature. For my present purpose the 'controlled' conditions are always much too far removed from the conditions in the natural habitat.

For numerous practical reasons such experiments under natural conditions begin with isolated, often seemingly insignificant behaviour components; their combined results form, together with descriptive-interpretative data, the raw material needed for the more comprehensive picture of each species we shall be aiming at.

A simple example which concerns a seemingly commonplace, at

first glance even trivial behaviour, is found in our own work on the function of the removal of empty egg shells by the Black-headed gull soon after a chick has hatched (Tinbergen, **87**, **89**). Not all bird species have this response, but in those that have it, it is usually both universal and prompt. In the Black-headed gull it seemed trivial because it takes each parent no more than some 10 sec each year. But because we had seen that even a few seconds' exposure of the brood was sometimes penalised by a 'dash-and-grab' attack of certain predators we had to assume that the response must have a positive advantage that overrides its negative effect. Comparison with other species showed that the response is most pronounced in species with camouflaged broods, and we therefore suspected that the conspicuous white rim and inside of the empty shell might well attract predators that hunt by sight.

The basic experiment consisted of laying out, scattered in the hunting area of such predators, equal sets of: (1) eggs which were given an empty egg shell at 10 cm distance, and (2) eggs without such a neighbour. In this way we mimicked the effects (on as yet unhatched eggs) of failure-to-remove and the effects of removal, i.e. of the behaviour of deficient and of normal parental behaviour. In a series of such tests the survival rates of the two types of 'broods' were compared.

The mortality by predation of the eggs-with-extra-shells turned out to be much higher than that of eggs not accompanied by an empty shell. The response therefore is a component of the protection of the brood by camouflage.

While it would be beyond the scope of this article to attempt a systematic review of the evidence about the adaptedness of single behavioural traits that can be found scattered in the literature, I will mention a few more examples if only to call attention to the variety of such traits of which the 'functional sense' is now becoming clear. The subject is in great need of a more comprehensive treatment. The functions of the 'dance' of worker honeybees (von Frisch, **30**); of echolocation in bats (Dijkgraaf, **23**; Griffin, **32**); of a wide variety of social signals, are already widely known, but a few more recent examples deserve brief mention.

The calls of many social birds show striking inter-individual differences; it has been shown for several species that this ensures individual interaction between parents and their own offspring (White, **94**; Tschanz, **90**).

In tropical seas one finds small fish that feed by cleaning the skin of other fish. They elicit appropriate behaviour in their 'hosts' by a combination of distinctive bright colours and a dance-like movement.

This symbiosis has given rise to a remarkable form of mimicry: a predatory fish of quite a different family mimics not only body shape and colour but also the peculiar dance of the cleaner fish. In this way it manages to deceive the 'cleanee' species and can prey on their live tissues (Eibl-Eibesfeldt, **26**).

Other studies concern more subtle aspects of behaviour. In times of abundance, many animals hoard food. In the fox it has been shown that such hoards are utilised, but only when reduction in the food supply demands it. A study of the ways in which hedgehogs rob fox caches revealed the survival value of 'scatter hoarding'—the fox consistently avoids 'putting all its eggs in one basket' (Tinbergen, **84**).

Among the many mating systems found in the animal kingdom, consistent monogamy, usual in species where both male and female care for the brood, is of particular interest. In some higher animals, even forms that reproduce seasonally, the same partners come or remain together for a number of years. In the Kittiwake, Coulson (**12**) has found that birds that do so achieve a higher breeding success than birds that take a new partner, as happens when the mate of the previous season dies. The partners obviously profit from 'running in' with respect to subtle aspects of collaboration.

HOW OUTSIDE PRESSURES ARE MET

One's horizon is considerably widened when one approaches the issue from the side of environmental pressures. For this one has first to learn to see the environment, so to speak, from the point of view of the animal—to distinguish '*Umgebung*' from '*Umwelt*' (von Uexküll, **92**). Of particular interest are those studies in which the pressure is exerted by other organisms, for instance predators, because here the pressures are themselves behavioural; both opponents in the ensuing 'battle of wits' employs sophisticated behaviour patterns. Again, a few examples must suffice.

The study of visual camouflage, i.e. adaptations aiming at not being detected, has long concentrated on structural aspects (Cott, **11**; Kettlewell, **43**, **44**). Step by step equally vital behavioural correlates are being discovered. Of these, egg shell removal has already been mentioned. The general features, developed convergently and with the use of widely different mechanisms by almost all camouflaged species are: staying motionless when required; selecting a matching substrate; taking up the least conspicuous position on this substrate; and, as recent work begins to show, living more scattered than the more tangible vital resources such as food would require (Tinbergen, Impekoven and Franck, **88**). The environmental pressure that all these species seem to have in common is a feeding method found in some predators that 'hunt by searching image' (Tinbergen, **81**; Croze, **16**). When such a predator stumbles upon one particular prey animal, however well camouflaged, it not only tends to continue its search in the vicinity of its first find, but also to concentrate its 'attention' on this particular type of prey, and to ignore many others. This type of feeding behaviour is now being analysed in more detail (Smith, **75**).

For another example I return to the Black-headed gull. Once our work on egg shell removal had called our attention to predation, the approach that concentrates on an outside pressure was also applied: in what ways do all the predators together threaten this species, and how does it defend itself? The main contributions come from the work of Kruuk (**47**) and Patterson (**7**), who made 'functional sense' of some seemingly commonplace, but none the less highly adapted features. First of all, the start of egg laying proved to be more highly synchronised within the colony than would seem to be required by seasonal 'peaking' of food supplies. Patterson found that both early and late broods were much more penalised by predation than peak broods. The interpretation is that synchronisation reduces the time of abundance for the predators—they are for a short period 'swamped by numbers', and as a consequence more broods escape them.

The spatial distribution of nests is likewise non-random. On the one hand the birds nest in colonies. It was found that the success of more isolated outlying nests and even of fringe nests was much lower than that of nests in the centre of the colony, and here too the penalty was due to predators: the members of the colony attack some predators in mass and so repel them—attacks by single pairs are far less effective. But another, contrary, pressure is exerted by predators that are not deterred by the mass attacks. For this, the gulls have a different defence. Within the colony the smaller inter-nest distances are under-represented—territorial behaviour ensures a certain degree of distance-keeping. This, together with the fairly good camouflage of the broods, gives a certain degree of protection against such predators (including neighbouring gulls) that do enter the colony; the indications are that they are hampered in their search by the spacing of the broods.

Even the obvious anti-predator devices of attack on the one hand and escape on the other show remarkable refinements. Kruuk compared the risk that each species of predator constitutes to the broods with the risk it embodies for the adults, and found that the balance between attacking the predator (which protects the brood) and escaping from it (which protects the adults) varies in accordance with the size, and with the hunting methods of the predators—with the risks they represent.

The overall result of such probing into, on the one hand the function of single behaviours, and on the other the ways in which single pressures are met, is the gradual unveiling of a fascinating, and even at this early stage already impressive, picture of the details of adaptedness, of the 'functional sense' of more and more aspects of behaviour that one might so easily have taken for granted.

One of the general lessons one learns in such work is that, seen in its overall context, no single adapted feature is ever ideal: the difference between perfection and adaptedness—which is no more than sufficiency (e.g. see Cain, **7**)—becomes clearer and clearer. This makes one wary of the claim, heard so often, that organisms show definite 'construction defects'. There are at least two reasons for caution:

1. The student of function meets time and again with evidence, such as found in the spacing of broods of gulls mentioned above, which shows that in order to survive, an organism, having to meet many different and often conflicting demands, has to compromise.

2. As has often been pointed out, each species is hampered by evolutionary inertia (Mayr, **58**)—this is another way of saying that it has to compromise between past adaptedness and present demands. The functional ethologist feels that the claim that 'we could make a better animal' reveals a lack of awareness of the extent of our ignorance.

FUNCTION AND CAUSATION

It is perhaps not superfluous to say that my emphasis on the value of studies of 'survival value' is not a plea against the analysis of mechanisms. On the contrary, every biologist knows that awareness of what an animal achieves guides his attention time and again to initially overlooked problems of causation. One sees an improbable and advantageous effect—an achievement—and one is inspired to ask how that achievement is brought about. I have often heard Karl von Frisch say that it was the conviction that the colours of flowers 'could not be there for nothing' that made him investigate the responsiveness of insects to colours, and then to a host of other problems of reception. Roeder likes to tell how, after having analysed both the functions and the mechanisms of the thoracic ears of certain moths, he rashly pronounced at an outdoor cocktail party that hawkmoths could not evade bats because, lacking these thoracic ears, 'they could not hear'. At just that moment he jingled the keys in his pocket and saw a mass panic among the Oleander hawkmoths feeding in the garden; this led him to the discovery of a new type of sound-receptor (Roeder, Treat and Vandeberg, **9**). Similarly, as we have seen, the indications that scattered living of camouflaged animals are a defence against 'search image hunters' guide one's interest both to the mechanisms of overdispersion in prey species, and to 'attention switching' and other hidden aspects of the feeding mechanisms of the predators (Dawkins, **20**, **21**).

The study of egg shell removal illustrates not only how functional studies guide research in causation, but also that the two approaches are mutually inspiring. The discovery that failure to remove the empty egg shells was penalised by predators made us turn to the stimuli that control the gulls' responses to egg shells. We found, among other things, that the gulls distinguished an egg shell from an intact egg not by its being hollow, nor by its broken outline, but mainly by its showing a thin edge. This opened our eyes to a new functional problem: what prevents the gulls from removing a newly hatched chick together with its egg shell when it is still half inside? We presented egg shells with small pieces of lead concealed in the far end. These models made the gulls start the first part of their response chain and take the shell in their bills, but they dropped it as soon as they had assessed the weight.

Another example: when we tested the gulls' responses to egg shell dummies of different colours we found that the natural khaki and white were carried more readily than other colours, however conspicuous, but that green objects were removed only rarely. Since tests on the effect of shape had already shown us that even flat paper rectangles (which show a thin edge) were often removed, the adaptedness of this low response to green was obvious: if the response to green were not inhibited, the bird might easily remove the leaves of surrounding plants, with the dual disadvantage of making the adults spend too much time away from the brood and of reducing cover for the chicks. Incidentally, the low response to green need not merely be due to a low overall sensitivity of gulls to green (Delius and Thompson, **22**): the incubation responses, at least of Herring gulls, happens to be elicited more readily by green eggs than by eggs of other colours (Baerends, **1**).

It is of course this continuous alternation between the search for survival value and the analysis of mechanisms that gives biology, at every level of integration, its particular flavour. The discovery of each particular achievement inspires one to find out 'how it is done';

conversely, the student of mechanisms derives satisfaction from understanding how the achievements of these mechanisms contribute to the animal's success.

COMPARATIVE ETHOLOGY

For our present purpose we have now to go one step further and try and place all these fragments of evidence in a wider, in essence evolutionary context. By combining the methods I discussed, and by making use not only of 'hard' evidence but also of 'inspired guessing', we can already now begin to sketch, however tentatively, the 'functional architecture' of the overall behaviour equipment of different species. For this the ethologist turns to the time-honoured comparative method: he studies the functional significance both of differences and of similarities between species—more precisely, he interprets differences between allied species in terms of adaptive evolutionary divergence; and similarities between otherwise dissimilar species in terms of adaptive convergence. This is applying to behaviour an old method which has been so extremely fertile in the past for our understanding of the evolution of structure—a lesson which, unfortunately, is in danger of being forgotten.

The study of behavioural divergence can be illustrated by E. Cullen's paper (**17**) which interprets functionally many peculiarities of a relatively aberrant gull, the Kittiwake. Unlike other gulls, Kittiwakes build a solid foundation platform for their nests before building the nest proper; their young hardly ever move from the spot; on their breeding ledges the adults are remarkably tame; when flapping their wings before they can fly the young do not jump about like other gulls; during copulation the female squats; during struggles over food the young whose share is being contested by its nestmates does not run away but merely faces away—these are only a few examples of the aberrant behaviour of this species, and at first glance they seem not only trivial, but also functionally puzzling. But Cullen pointed out that these, and many more, details made excellent functional sense when the oceanic habitat, and the correlated habit of breeding on extremely narrow ledges on vertical cliffs, were seen as the essence of their particular niche. This is an anti-predator device: on their cliffs Kittiwakes are much safer from predation than are most other gulls, who breed in much more accessible habitats.

On such ledges, nests require a broad and horizontal base; the young must have behaviour that protects them from falling down; the adults not only can afford to be tame on the cliff, but the reduced need for fleeing or attacking frees time for other, at times vital pursuits. The extreme intraspecific aggressiveness of the species, its shyness when collecting nest material; the habit of robbing nest material and the countermeasure of guarding the nest much more carefully than other gulls even before it contains eggs are other characteristics that fit into this picture. Some of these traits have been developed convergently by other species which breed in similar situations (Cullen and Ashmole, **18**; Hailman, **33**; Smith, **76**).

Studies of this type reveal, first that the evolutionary radiation has been, down to most surprising details, adaptive, and secondly, that the adaptedness pervades the entire system of a species—that the overall behaviour equipment of each species forms a closely woven web. It is worth stressing that even the nature of a species' social behaviour is indirectly influenced by the peculiarities of the breeding habit, by feeding habits, and by predator pressure—a trend emerging from studies of other animals as well (Crook, **14**).

Here again, attention to functional aspects is leading to more detailed analysis of the causation of behaviour. Inspired by Cullen's work, McLannahan (**55**) investigated how Kittiwake chicks manage to avoid falling off the cliff. Her main findings were that Kittiwake chicks were no less mobile *per se* than those of other gulls, but they stay on the ledge (1) by being more strongly attached to the parents (who also contribute by spending more time on the nest than other gulls), (2) by being attracted by the cliff wall, and above all (3) by having, from the moment of hatching, a near-perfect abyss-avoiding response. The latter response shows interesting details, and several of these are now being studied comparatively in other gulls. As we could by now expect, a clear parallel is being found between the development of the response and the requirements of the breeding habitat (J. Tinbergen, pers. comm.).

My next example centres on feeding rather than on predator pressure. Wynne-Edwards (**96**) has called attention to the phenomenon of deferred maturity and interpreted its function as an altruistic brake on population growth. One example is the Oystercatcher, a wader which normally does not breed until it is 4, occasionally when 3, years old. When the feeding behaviour of this species was studied in detail (Norton-Griffiths, **64, 65**), its remarkable skill in opening bivalves such as mussels and cockles and other hard-shelled invertebrates of the tidal zone was revealed. Most members of the species rely largely on this locally superabundant food supply, which is only occasionally available to less specialised feeders. Many individual Oystercatchers even specialise on one particular food species, and, more surprising still, on a particular method of opening the shells.

Some mussel specialists gain access to the flesh by stabbing between the shells from the dorsal side, cutting and tearing the adductor. Others proceed by hammering a hole in the shell's ventral side and so damaging the adductor. 'Hammerers' and 'stabbers' follow the same routine. First, the shells are opened more widely by sideways levering movements of the bill. Then the bird turns at right angles to the mussel's axis and forcefully prises the shells even wider apart. It then separates the flesh from the shells by a series of rapid chiselling movements, shakes the entire mussel loose and swallows it. The whole operation lasts approximately 30 seconds.

This remarkable specialisation has a number of interesting functional implications. First of all, the chicks could not possibly fend for themselves, as those of most other waders do. Even if they were to have the skill, they lack the required instrument, the long and exceptionally strong bill. This makes it functionally understandable that Oystercatchers, as exceptions among waders, feed their young. They do this not only until the bills of the young are fully developed, but for a much longer period. This is adjusted to the slow development of the feeding skill of the young. The development begins early. Already in the first days after hatching the chicks show, when near-satiated, the incipient, recognisable precursors of the adult feeding movements. These are at first useless for feeding, but acquire their ultimate function through at least three learning processes. First, as a consequence of their parents' concentration on one type of food, the young of, for instance, mussel-eaters become conditioned to mussel shells as containers of food. Secondly, the young improve by practice their skill in extracting the flesh from the shells. Thirdly, and most surprisingly, it is parental example that decides whether a young bird shall become a hammerer or a stabber. All young have the potential to develop either technique, but it is only the parents' technique that they do develop. This incidentally is an example of 'cultural' transfer, yet it is not cultural evolution.

The task of feeding a brood is so demanding that Oystercatchers show a steep drop in weight during the breeding season (Dare, **19**). The development of the skill by practice proceeds so slowly that, while a young bird is within months able to feed itself, it is not until it is 3 or 4 years old that it can feed efficiently enough to reach the weight necessary to sustain the loss inflicted by raising a brood. It is therefore clear that the Oystercatcher pays for the evolution of its remarkable feeding skill by having to feed its young, and by having to go through a long period of learning this skill. Since Oystercatchers are long-lived they can afford to breed late: individual birds who defer breeding may well in the end raise more young than birds that overstrain themselves by breeding earlier. The interpretation of deferred breeding as an altruistic trait, which would require explanation in terms of group selection, is clearly superfluous for the Oystercatcher. However, for our present purpose it is sufficient to focus attention on the functional interrelationships between the several peculiarities of this specialised wader; on the many repercussions of having opened up, by means of a special technique, an abundant source of food.

These few examples demonstrate how the systematic study of adaptedness begins to reveal, in steadily growing detail, not only how precarious are the 'balancing acts' which animals have to perform, but also how intricately adapted is the behaviour of each species to its own particular niche. Or, to express it in another way: studies of this type begin to reveal *in concreto* the details of what natural selection has produced. And, as I hope to show presently, it is this concrete knowledge we shall need.

Adjustability

Turning to my second main topic, that of the extent of individual adjustability: this is an aspect of the old 'nature-nurture' problem: to what extent is behaviour genetically programmed and to what extent is it further improved by individual modification? I approach the problem from the angle of phenotypic adjustability—its extent and its possible limitations—because this is the practical problem facing Man.

At a time when 'learning' was overstressed, the reaction of ethologists against the '*tabula rasa*' concept of behaviour development was an extremely useful contribution. But the dichotomous classification into innate and learnt behaviour has rather outlived its usefulness for what behaviour students are now, belatedly, doing: analysing the developmental process (Lehrman, **52**). They begin to discover, (1) by raising animals in different surroundings and (2) by interfering with internal development (done far less often because it is so much more difficult) that, even though the details of the development are extremely complex and insufficiently known, behaviour patterns can be placed, according to their development, on a scale ranging from highly resistant to variations in the environment to highly modifiable. The fact, demonstrated by the ethologists of the thirties, that many behaviour patterns develop almost perfectly in either grossly deprived situations or even against contrary environmental pressures is now hardly worth stressing (Lorenz, **53**; Hinde, **37**). What is relevant to my subject is the fact, now becoming increasingly clear, that learning is not random, but is often a highly selective type of interaction with the environment.

The work of Thorpe (**78**), Marler (**57**), Konishi and Nottebohm (**45**) and others on the development of 'song' in some passerine birds can be taken as an example. Song is an elaborate and distinctive motor pattern which functions as a signal in territorial behaviour and mating. When males of the chaffinch are reared without being allowed to hear the song of an adult male, they do not develop the full adult song, but produce a less elaborate 'warble', which has a few but not all the characteristics of the normal song. For the full song to develop young birds must hear, at an early stage, the song of an experienced adult (which, however, they do not reproduce until much later). Such particularly sensitive periods for learning are now known in many instances. In the Chaffinch it has further been shown by Thorpe that not all 'teacher' songs are learnt with equal readiness—the birds are biased (i.e. pre-programmed) in favour of learning those songs that show certain characteristics of the natural song of their species.

The work on the Oystercatcher mentioned above provides another clear example of internal control of learning. Young Oystercatchers are conditioned to the type of prey that the parents provide but conditioning to, for example, a mussel shell does not happen in chicks that are merely fed with blobs of mussel flesh in opened shells—it is only those young that receive food, *and* have to chisel it loose from the shell that become conditioned. These are only a few of the many examples accumulating (see, for example, Seligman, **73**; Hinde and Hinde, **36**) that show that the genetic instructions for the development of behaviour include instructions for phenotypic adaptation—that even learning is not random, but its occurrence, what is learnt and how it is learnt, are prescribed internally within relatively narrow limits, and in addition that these prescriptions are different in different species—each of them is 'programmed for learning' in its own, and adapted way. This expresses itself not only in limitations of what is learnt, but also in a more positive way. A special example of this is exploratory behaviour. This can only be described as behaviour which has the function of creating the opportunities for individual programming. In control of its motivation in its sensitivity for very special external conditions, in its cessation when the environment has become 'familiar', i.e. as soon as salient aspects have been added to the animal's 'knowledge', it is a very beautifully adapted behaviour—adapted to the need to create opportunities for relevant phenotypic adaptation—for maximum success.

An extreme form of exploratory behaviour, known as 'locality study', is shown by some insects. Best known is its application by solitary *Hymenoptera* in 'homing' to the nest site (Tinbergen, **82**, **86**). Manning (**56**) found that foraging bumblebees learn the position of some individual plants by means of a locality study. But he found, in addition, that such a locality study is made after the discovery of a new *Hypoglossum* plant with a rich nectar supply, but not after the discovery of a new Foxglove. When returning to plants of the latter, of which the flowers are visible from a much larger distance, the bumblebee relies on its roaming flight over a large, known area, and on seeing the flower spikes from a distance. Whether or not it makes a locality study therefore depends on surprisingly detailed aspects of the situation.

Resisting the temptation to go into more detail, I want to stress the important point that even where adaptedness depends to a large extent on learning, the modifications are themselves internally, ultimately genetically controlled; they are an individual continuation of the process of adaptation; and this supplementary programming varies from species to species; within a species it varies from one developmental stage to the next; from one behaviour system to another; and even from one situation to another. These aspects of the internal control of modifiability are relevant to our own species as well.

Ancestral Man

What lessons can we draw for our own species from these probes into functional ethology? Man is being studied by so many specialised disciplines that it might seem almost impertinent for animal ethologists to put a word in. But the fragmented state of the human sciences justifies and even requires the mobilisation of as many relevant disciplines as possible. Ethologists begin to believe that at least some of the methods developed in their science could, if both their power and their limitations are borne in mind, profitably be applied to some important human problems.

At the start I have to emphasise the difficulty to applying to modern Man the same functionally oriented, comparative method as we are applying to other animals. This method works well with products of genetic evolution, which still live in the environment to which natural selection has adapted them, and whose behaviour as a consequence is to a large extent constant throughout the species. But because both our behaviour and our environment have changed so much since the cultural evolution began to gather momentum, we are faced with a bewildering variety of anthropogenic modifications—one could say distortions—of environments, and of behaviour systems. Before we could apply the comparative method, we would have to 'peel off' these cultural variations, and reconstruct the behaviour and the environment of our precultural ancestors. Not until we can perform this dual task of reconstruction of environment and behaviour could we sketch a picture of the genetic adaptedness on which our cultural evolution has been superimposed, and which, conversely, has influenced the directions it has taken. To perform this reconstruction, and then to apply the comparative method to Man as well, is what the often, but in many respects unjustly, criticised 'Naked-Apery' is attempting to do. It is a historical exercise, with all its inherent uncertainties.

We do know a little about the environment of early Man. Initially derived from a forest-dwelling 'swinger' he has become a bipedal inhabitant of a more open habitat, one richly provided with an under-exploited food supply. Whether or not Hardy's imaginative idea (**34**) will receive confirmation that a semi-aquatic phase has also been involved, can be left open, but there seems little doubt that our immediate ancestors have occupied a terrestrial niche—which, as Schaller and Lowther (**72**) have so convincingly argued (see also Schaller, **70**, **71**) early *Homo* could have invaded without much competition.

For the reconstruction of early human behaviour it is only to a very limited extent possible to draw on fossil evidence. It does reveal the early appearance of, for example, bipedal locomotion and a switch from a vegetarian diet to that of a hunter-gatherer; and early cultural developments such as tool-making and the use of fire. But beyond this we must rely on other methods of reconstruction. These are all based on the conviction that cultural evolution has not been random, but that it has affected principally those aspects of our ancestral behaviour equipment that were relying most on individual programming, and has changed to a far lesser extent the more internally programmed, more resistant traits, which therefore must be taken to reflect most clearly our ancestral heritage. Evidence of such environment-resistant traits can be expected to come mainly from two sources. Whatever is least variable between cultures, and whatever is least variable within a culture and appears even in spite of environmental pressures in a culturally modified society, is most likely to reveal an ancient, environment-resistant 'deeper structure'.

It has been pointed out by Morris (**62**), and in my opinion with justification, that anthropology and ethnology have until recently tended to concentrate more on differences between cultures than, as would be required for our purpose, similarities. As an example of a programme that aims specifically at the study of such similarities I mention recent work of Eibl-Eibesfeldt (**28**). It would be far beyond the scope of this paper, and certainly beyond my competence, to try and sketch what intracultural analysis is revealing about the deeper structure of our modern Western behaviour. The sciences most directly confronted with resistant phenomena within our culture, the psychopathological and the educational sciences are involved in a process of conceptual and semantic fermentation and are in addition split up in innumerable schools. It would be an extremely difficult but also a very important exercise indeed to cut through the barriers that separate such sciences as palaeontology, archaeology, anthropology, normal and abnormal psychology of both adults and children, and to extract the already available and quite considerable evidence that is relevant to our topic. This would also help to guide future research towards a better understanding of the 'deeper structure', the ancient roots of human behaviour. In this programme, some methods developed in ethology could be of great help, as can be seen from the collection of studies published recently under the editorship of Blurton Jones (**4**).

Yet, in spite of the lack of a unified approach, and notwithstanding the fragmented evidence, it is already possible to make a tentative, inspired guess at the ancestral behaviour equipment of our species—a kind of thumbnail sketch. As we go along we can check, as I did with the Kittiwake and the Oystercatcher, whether such a sketch would make functional sense, as of course it should if we are guessing in the right direction.

Our reconstruction can best start from an assumption that few can doubt, namely that our bipedal locomotion, the prolonged helplessness of the human infant and its need for extended parental care, as well as our pronounced sexual dimorphism are old, hardly modified characters. Cross-cultural as well as archaeological-palaeontological evidence suggests that early Man lived in relatively small groups—so small that, as in many other Primate societies, all individuals must have known each other personally. There is also no doubt that early Man has been more of a hunter than his close relatives, certainly more than the surviving apes (Lee and DeVore, **51**). The well-established fact that early man was able to kill animals much larger than himself suggests strongly that hunting was, at least on occasion, done in groups. In this respect comparative evidence on other mammalian hunters strengthens our reconstruction: some members of the dog family; the Spotted hyaena and, among the cats, the Lion hunt socially, and this allows them to live in part on much larger prey than their solitary relatives (Mech, **59**, **60**; Schaller, **70**, **71**;

Kruuk, **48**). In Man, the hunting of the larger animals must have been done by the physically stronger males, since the adult females, even though they could carry their infants around, had to be more tied to a secure base. (In this respect our dimorphic species differs from, for instance, wolves.) In our hunting groups there must have been collaboration, and comparative as well as intracultural evidence suggests that this must have been based on a dominance order—a phenomenon incidentally that entails much more than the word dominance suggests; lower ranking animals do not simply fear their superiors, they also 'respect' them, follow their leadership, and learn from their example (Chance, **8**; Kummer, **49**). In Man, the male is not only a hunter, but also a provider (Washburn and Lancaster, **93**) and to a certain extent an educator; this makes it likely that the nuclear family group has at an early stage included the father as well as the mother. Among animals, those species in which both male and female take part in the care of the young have evolved monogamy, and a long-lasting pair-bond with its accessory of falling in love. If, as seems likely, early Man was also monogamous (with perhaps incidental bi- or polygamy), the hunting mode of life created a special problem. Hunting large animals (who have large ranges) required long hunting trips and therefore long absences. This would require strong pair-bonding devices. As such, monogamous animal species use various forms of 'extraneous' behaviour systems, such as joint nest building, feeding of the female by the male, mutual preening or grooming, etc. In Man, this is where sexual behaviour seems to come in. Coition between partners is recognised as having a strong bond-reinforcing function. In this context the fact that the readiness to mate is far less cyclical in Man than in other species should be considered. It is also significant that the use of sexual behaviour or parts of it for other purposes than mere fertilisation is widespread among Primates (Wickler, **95**). In the present Primates it is used mainly as a ritualised signal to stabilise the dominance hierarchy: the male mounting act signals superiority, the female posture inferiority, even in non-reproductive encounters among individuals of the same sex. Sex behaviour was therefore so to speak already available for secondary non-reproductive functions. Comparative studies have shown that such dual use of behaviour patterns for both a primary and a secondary function is widespread in the animal kingdom. To the comparative ethologist, the condemnation of 'sex for pleasure' in marital and pre-marital context alike seems to reveal a lack of biological knowledge; it also ignores the realities of married life even in modern society.

The young of our species had, and have a relatively poor non-learnt behaviour repertoire (though undoubtedly richer than has often been assumed (Blurton Jones, **4**; McGrew, **54**)). The matrix of movements, signals, sensitivity to signals and motivations is, more than in any other species, improved phenotypically. This is ensured not only by social interaction—at first with the mother, then with peers, then with an even wider circle—but also by the young's own, extremely important, exploratory behaviour. As in other Primates, this exploratory behaviour blossoms only in the security of maternal supervision, and later in that provided by other friendly individuals. The long development culminates in full incorporation into the adult society, the result not only of relaxation of parent—infant ties but also a active self-assertion by the adolescents—incidentally, the basis for a 'generation gap' which, under the conditions of a vastly accelerated cultural evolution, has now created the need for a *mutual* adjustment between adult and adolescent instead of a mere waiting for the young to conform. Comparisons with group-living Primates and Wolves, hyaenas and Lions, as well as cross- and intra-cultural evidence in Man render it further probable that inter-group hostility, particularly between males, and intra-male group friendships are likewise old characteristics. The gist of L. Tiger's interpretations in *Men in groups* (**79**) seems to me to be biologically sound. Whether or not this population structure was accompanied by group territories and consequent territorial inter-group hostility can be left open, although in view of what we know of the importance of intricate knowledge of the hunting area, and of the strong tendency of men in many cultures to behave like group-territorial animals, this would seem to be very likely.

It is not necessary here to work out this reconstruction in more detail. As I have said before, it is my belief that it is in outline sound, and also that by a more systematic and more purposeful collaboration of the many relevant sciences, it can be better substantiated, and also elaborated in much more detail.

DISADAPTATION AND RE-ADAPTATION

Already at this stage it is possible to see a little more clearly how drastically and in what respects our new environment differs most from the precultural habitat. Urbanisation is perhaps the most striking development. It has carried with it not only crowding, but the formation of very large and in particular anonymous societies, very far removed from the small in-groups of early Man. We are also submitted to an enormously increased quantity of input, not only in the form of general, amorphous sensory input in the auditory and visual sphere, but also in the form of quantity of communication through the mass media. The work of countless industrial workers has become extremely monotonous and very far removed from the meaningful and immediately satisfying occupations of the individual craftsman, and certainly from those of the primitive hunter and the primitive agriculturalist. The education of children has changed almost beyond recognition into an extremely demanding training for modern citizenship.

The question that faces the comparative ethologist is: are there signs that this new situation imposes demands on 'human nature' that exceed the limits of its phenotypic adjustability? Are there intolerable pressures, and are there, conversely, gaps, pockets of missing outlets for behaviour patterns that have strong, perhaps compulsive internal determinants? Ethologists believe that there are such signs, and I select three of them for a brief discussion.

Shortly after World War II, J. Bowlby (**5**, **6**) traced back certain disturbances of social behaviour to disruptions of the early phases of affiliation, of bonding between mother and child. Bowlby saw straightforward deprivation of the presence of the mother during longer or shorter periods as the primary cause of a failure in children, first to form personal bonds with the mother, and subsequently of social bonding of any kind—he argued that socialisation comes about by a widening of the circle of friends which is only possible if the first personal bond is successfully established. He makes it clear that the young child needs a stable, loving mother or substitute mother, and that modern social conditions often disrupt or even entirely fail to provide for this early phase of socialisation. Work on the development of mother–infant relations and of socialisation in other mammals, in part inspired by Bowlby's work, gives increasing support to his thesis. Indeed, it becomes very likely that it is not just the presence of a stable mother figure, but an extremely intricate pattern of maternal behaviour that is required. Even very mildly disturbed mothers, such as slightly insecure, or slightly preoccupied, working mothers may unintentionally deprive their children. Conversely, over-intrusive, underoccupied mothers may well, through interfering at moments when a child wants to play on its own or with peers, make a child withdraw. There are further signs that, as in other mammals, the affiliation may have to start immediately after birth, and the importance of the ethological studies of mother–infant interactions that are now being made (Harlow and Zimmermann, **35**; Spencer-Booth and Hinde, **77**) can hardly be overstressed. The paucity of knowledge of these early phases of human life is astonishing, and so, incidentally, is the assertiveness of many theorisers. A number of incidental, at first glance seemingly disconnected, observations of family life in Man and other mammals, and also in birds, suggest to me that such studies may well reveal even more widespread damage to socialisation than Bowlby has pointed out.

As another possible sign of behavioural stress I should like to mention briefly the disorder, or perhaps group of disorders, now generally called Early Childhood Autism, or Kanner's syndrome (Kanner, **42**; Bettelheim, **3**; O'Gorman, **66**; Tinbergen and Tinbergen, **80**). There are indications that the recent widespread interest in this serious aberration is due to a real increase of incidence rather than to belated recognition (while its discoverer Leo Kanner still found it difficult, in 1943, to obtain sufficient information, there are now, in Britain alone, some 6,000 children officially diagnosed as autistic). The syndrome is characterised by a very nearly total lack of socialisation, by complete withdrawal from, and even violent rejection of other persons, and by underdevelopment of speech, and a number of other skills. A considerable number of autists are damaged for life.

Together with my wife I have compared some aspects of social and socially determined behaviour of normal and autistic children, applying methods developed largely in ethology for the analysis of the motivation underlying non-verbal 'expressions of emotions'. We could confirm and elaborate the evidence which had earlier led Hutt and Hutt (**40**) to state that many normal children can on occasion show all the components of Kanner's syndrome. By analysing the stituations in which this occurs and by studying the forms of therapy which appear to have success, we arrived at the following, tentative conclusions.

1. The distinction between normal and autistic children is far from sharp, and a considerable number of 'normal' children may well be mildly autistic.

2. Motivational analysis indicates that both temporary and permanent autists live in a state of motivational conflict between hyperanxiety and, as a consequence, frustrated sociality.

3. This conflict can become so severe that the child withdraws from and rejects not only strange persons and environments, but also those that to a normal child become familiar: this can lead to a rejection even of its closest relatives.

4. As a result, socialisation is severely hampered, in fact resisted by these children, and as a consequence learning processes that normally form part of socialisation, such as the acquisition of overt speech, exploratory behaviour, and certainly learning by social instruction are likewise impaired.

5. Whereas naturally (i.e. either genetically or 'organically' deviant) timid children are most likely to develop the syndrome, it may well be caused to a much larger extent than is generally recognised by shortcomings in the social environment, to be found particularly in family life, which in certain strata of urbanised society shows certain disruptions.

In other words we believe that at least some forms of autism are the consequence, and indications of certain forms of increased social stress, and that autism may well be an 'early warning' of harmful effects of the cultural evolution. The phenomenon is certainly in great need of further study, and ethological methods may well help remove the uncertainty and disagreements about its causation, and so ultimately assist in reducing its impact.

Next to these two problems concerning child-rearing and family life, the cultural changes that have imperceptively, but none the less drastically, influenced educational practices are of at least equal importance. Most of us take institutionalised education for granted, and for preparing children for the highly specialised parts they will have to play in the modern anthropogenic environment, schooling of some sort is of course indispensable—to think of abolishing schools, as some do, seems totally unrealistic. But it must not be forgotten that schools are a relatively recent cultural phenomenon, and that as such they have to grow with the times. To the ethologist it is clear that we will have to do some hard thinking about both the aims and the methods of education lest we increase, in this sphere too, social stress to beyond what is tolerable.

Socially of course school is a good thing because it brings peers together. This would seem to be in harmony with the 'deeper structure' of human social life. It is also a healthy antidote to the isolation that the need for small families and, paradoxically, urbanised living the crowded but socially lonely conditions of high-rise flats, are threatening to force upon us.

But when we compare the educational activities of present-day schools, and the progressive extension of compulsory schooling, with the educational system of contemporary 'primitive' societies, and, by inference, of ancestral Man, a few questions arise that deserve much more serious consideration than they are being given, for the differences are much more striking than is generally realised.

The rapid growth of technology in the widest sense requires of course that we prepare each generation for playing their part in a society of increasing complexity. But this very speed of change also carries with it the need for each new generation to fit into quite a different, and far more complicated society than that of its parents. Each generation has to learn a great deal more than the previous one. Even we ourselves have seen, in our own life span, a vast quantitative and qualitative increase of what has to be learnt. And this intensified and extended programme has to be met by children who live in a climate of already increased stress and overall input. Modern conditions force us to raise a generation that is at the same time more knowledgeable and optimally flexible and adjustable. Do our educational practices meet these new and perhaps contradictory demands? The question has often been asked, and many educationalists have voiced doubts. In British infant schools and primary schools a great and beneficial revolution is in progress, but even more far-reaching changes may well be needed. Among the many educational innovators and would-be innovators who are expressing opinions on this issue, we might do well to pay most attention to those who base their proposals on close observation on how children learn, and how they fail. With Maria Montessori still as one of the great modern pioneers, most of these students of child development have stressed the need for less imbibing of 'knowledge' and more 'self activity'.

To the comparative ethologist, this seems to make eminent functional sense. In 'primitive' societies, presumably in this respect more similar to ancestral communities, learning depends partly on exploratory 'play', partly on social imitation, and only partly on deliberate instruction by adults. It seems to me that we have disproportionally increased the part played by social instruction, and that in doing so we are likely to hamper, indeed to stunt and distort development in two ways: we are quite possibly *suppressing* exploratory learning; and we are undoubtedly calling up serious *resistances* against social instruction. This merits a little elaboration.

It is one of the valuable characteristics of our species that the tendency to explore, the sense of curiosity, continues for much longer in the life of the individual than even in the highest Primates. Not only the specialised tasks of scientists and technologists, but many other activities in modern society require open-mindedness and an imaginative, exploratory attitude.

The conditions under which exploratory learning flourishes are security, a minimum amount of interference by adults (as distinct from guarding), time and opportunity, and an environment which invites exploration.

I cannot resist relating one little incident which, although one would hardly expect this from the literature on child development, is in my experience representative. A 12-month-old boy, guarded by his aunt and his grandmother, was observed crawling about over a sandy slope which was bare but for isolated rosettes of ragwort and occasional thistle plants. After having moved over many ragwort rosettes without showing any reaction to them, he happened to crawl over a thistle, whose prickly leaves slightly scratched his foot. Giving a barely perceptible start, he crawled on at first, but stopped a second or so later, and looked back over his shoulder. Then, moving slightly back, he rubbed his foot once more over the thistle. Next he turned to the plant, looked at it with intense concentration and moved his hand back and forth over it. This was followed by a perfect control experiment: he looked round, selected a ragwort rosette and touched that in the same way. After this he touched the thistle once more, and only then did he continue his journey. To ethologists this is only one of many examples of true experimentation in a pre-verbal child; of highly sophisticated exploration.

Yet to all observers of children it must be striking how early their exploratory interest wanes when their school training gets under way. Among the many educators who see a causal connexion between this waning and being 'drilled', the perceptive American educationalist John Holt (**38**, **39**) deserves particular attention because he has put his finger on two fundamental, yet not sufficiently acknowledged effects of school teaching, with its corollaries of continuous testing and giving good marks as rewards and low marks as punishments. He points out that a high proportion of children not only fail to respond, but actively defend themselves against this form of social instruction (and so defeat the teacher) by a mixture of bored rejection and deep-rooted apprehension, by fear of being found wanting. I am convinced that these points deserve close attention, because the tendency, already noticeable in primary school education, of teaching as much as possible by arousing and stimulating exploratory interest rather than by regimented instruction, is not only biologically sound but yields promising results. Nor is this surprising to those who have watched the complete absorption, the perseverance and the patience of children who 'work' in this way and the intense satisfaction they derive from it. This 'wind of change' is also touching secondary education, and will invade higher education, if only because students who have acquired a taste of joint and guided exploration rebel against 'being told', and already demand more exploratory forms of learning. It would seem to me that such a return to a biologically more balanced form of education could also lead to the raising of a type of person which our society needs now. For these needs have been rapidly changing. While a few generations ago we required above all competent professionals, the emphasis may in the future have to be rather on adjustability, openmindedness, ability to judge, ability to plan far ahead and similar qualities, not only in the leaders, but also in those by whose consent the leaders exert their influence.

My colleagues working in one of the many human sciences may well wonder what has given me the temerity to stray so far outside my home range. The obvious answer would seem that the future of our species is too important to be left to any one group of specialists. The human sciences are, as no one will deny, still very far from a unified discipline. In building such a discipline the collaboration of biologists will be necessary. And a functional ethologist, who is continuously faced with the precariousness of survival in animals, is in a position to see at least some aspects of Man's unique position that may not strike students of Man as such—and he is extremely alarmed by what he sees.

As I have argued, it is the comparison of the adaptedness of animals with that of Man which reveals that our conquest of the environment is causing habitat changes at a pace that genetic evolution cannot possibly match—not even if directed and speeded up by the, morally and practically doubtful, genetic engineering that some have proposed. In order to retain adaptedness we shall have to rely on phenotypic adjustment, and one of the most relevant new insights is that the range of individual adjustability is severely limited, and yet that our educational practices may not exploit to the full the developmental potential that we do possess.

Opinions differ about the imminence of harmful stress phenomena. I have argued that the behaviour student considers it very well possible, indeed likely, that we are reaching a point where the 'viability gap'—the gap between what our new habitat requires us to do and what we are actually doing—is becoming so wide that our behavioural adjustability is already now being taxed to the limit. The conclusion seems inescapable that we shall soon be faced with a task of 'bio-engineering' for the purpose of restoring our adaptedness, or rather of re-establishing adaptedness at a new level. The three examples I have mentioned (damage through mother deprivation, autism, and effects of lop-sided teaching habits) may well indicate that this task is already now becoming urgent.

It will involve at the same time a restoration of a tolerable environment and the development in ourselves of the highest possible level of flexibility. It will be a task that will require all our resourcefulness, for, as I said, we have no precedent to go by. Briefly, it will amount to no less than *phenocopying, in a very short time, and without paying the tax of massive weeding-out of comparative failures, some-*

thing that has so far been produced only by genetic evolution, which had aeons of time, and which did pay the price of gigantic numbers of errors.

In this task we cannot possibly succeed unless we know *in concreto* the new pressures that we are creating, and how these pressures could be either met or reduced. And while functional ethology helps us in identifying these pressures, it will be the knowledge of behaviour mechanisms, and of mechanisms of behaviour development that will have to form the basis for whatever engineering will have to be undertaken.

The execution of such an engineering task may at the moment seem to belong in science fiction, but I am convinced that sooner or later it will become a political issue. Knowing what we do about political decision-making I believe that it will be useless to call upon people's altruism or use other arguments of a moral nature. Rather, the scientist will have to point out that the prevention of a breakdown, and the building of a new society is a matter of enlightened self-interest, of ensuring survival, health and happiness of the children and grandchildren of all of us—of people we know and love.

No one can say how soon science will be called upon for advice, but if and when that time comes, we shall have to be better prepared than we are now. The main purpose of my paper is therefore to urge all sciences concerned with the biology of Man to work for an integration of their many and diverse approaches, and to step up the pace of the building of a coherent, comprehensive science of Man. In this effort towards integration animal ethology cannot stand aside—indeed I for one believe that, provided it will be given the opportunity for further development, it can render invaluable services.

REFERENCES

1 BAERENDS, G. P. (1957). 'The ethological concept "releasing mechanisms" illustrated by a study of the stimuli eliciting egg retrieving in the Herring gull', *Anat. Rec.*, **128,** 518–19.

2 BAERENDS, G. P. and R. H. DRENT (eds.) (1970). 'The Herring gull and its egg', *Behaviour Suppl.*, **17**.

3 BETTELHEIM, B. (1967). *The Empty Fortress: Infantile Autism and the Birth of the Self.* London, Collier–Macmillan.

4 BLURTON JONES, N. (ed.) (1972). *Ethological Studies of Child Behaviour.* London, Cambridge University Press.

5 BOWLBY, J. (1951). *Maternal Care and Mental Health.* Geneva, W.H.O.; London, H.M.S.O.; New York, Columbia University Press.

6 BOWLBY, J. (1969). *Attachment and loss.* Vol. 1. *Attachment.* London, Hogarth Press.

7 CAIN, A. J. (1964). 'The perfection of animals', in *Viewpoints in Biology,* ed. J. D. Carthy and C. L. Duddington, **3,** 36–62.

8 CHANCE, M. R. A. (1967). 'Attention structure as the basis of primate rank orders', *Man*, n.s. **2,** 503–18.

9 COMMONER, B. (1966). *Science and Survival.* New York, Viking Compass Book.

10 COMMONER, B. (1972). *The Closing Circle.* London, Jonathan Cape.

11 COTT, H. B. (1940). *Adaptive Coloration in Animals.* London, Methuen.

12 COULSON, J. C. (1966). 'The influence of the pair bond and age on the breeding biology of the Kittiwake gull *Rissa tridactyla*', *J. Anim. Ecol.*, **35,** 269–79.

13 CROOK, J. H. (1965). 'The adaptive significance of avian social organisations', *Symp. zool. Soc., Lond.*, **14,** 181–218.

14 CROOK, J. H. (1970). 'Social organisation and the environment; aspects of contemporary social ethology', *Anim. Behav.*, **18,** 197–209.

15 CROOK, J. H. (1970). 'The socio-ecology of Primates', in *Social Behavior in Birds and Mammals*, ed. J. H. Crook, 103–66. London and New York, Academic Press.

16 CROZE, H. (1970). 'Searching image in Carrion crows', *Z. Tierpsychol., Suppl.*, **5,** 1–86. Berlin, Parey.

17 CULLEN, E. (1957). 'Adaptations in the kittiwake to cliff-nesting', *Ibis*, **90,** 71–87.

18 CULLEN, J. M. and N. P. ASHMOLE (1963). 'The Black Noddy *Anous tenuirostris* on Ascension Island. II. Behaviour', *Ibis*, **103*b***, 423–46.

19 DARE, P. J. (1966). 'The breeding and wintering population of the Oyster-catcher in the British Isles', *Fishery Invest., Lond.* (Ser. 2), **25,** 1–69.

20 DAWKINS, M. (1971). 'Perceptual changes in chicks: another look at the "search image" concept', *Anim. Behav.*, **19,** 566–74.

21 DAWKINS, M. (1971). 'Shifts of "attention" in chicks during feeding', *Anim. Behav.*, **19,** 575–82.

22 DELIUS, J. D. and G. THOMPSON (1970). 'Brightness dependence of colour preferences in Herring gull chicks', *Z. Tierpsychol.*, **27,** 842–9.

23 DIJKGRAAF, S. (1946). 'Die Sinneswelt der Fledermäuse', *Experientia*, **2,** 1–31.

24 DUBOS, R. (1968). *So Human an Animal.* New York, Scribner.

25 EHRLICH, P. R. (1968). *The Population Bomb.* New York, Ballantine.

26 EIBL-EIBESFELDT, I. (1959). 'Der Fisch *Aspidontus taeniatus* als Nachahmer des Putzers *Labroides dimidiatus*', *Z. Tierpsychol.*, **16,** 19–25.

27 EIBL-EIBESFELDT, I. (1967). *Grundriss der Vergleichenden Verhaltensforschung.* Munich, Piper.

28 EIBL-EIBESFELDT, I. (1972). *Die Ko Buschmann—Gesellschaft: Gruppenbildung und Aggressionskontrolle.* Munich, Piper.

29 FRICKE, H. (1971). 'Fische als Feinde tropischer Seeigel', *Marine Biol.*, **9,** 328–38.

30 FRISCH, K. VON (1967). *The Dance, Language and Orientation of Bees.* Cambridge, Mass., Harvard University Press.

31 GOLDSMITH, E. *et al.* (1972). 'A blueprint for survival', *The Ecologist*, **2,** 2–42.

32 GRIFFIN, D. R. (1958). *Listening in the Dark.* New Haven, Yale University Press.

33 HAILMAN, J. P. (1965). 'Cliff-nesting adaptations of the Galapagos Swallow-tailed gull', *Wilson Bull.*, **77,** 346–62.

34 HARDY, A. C. (1960). 'Was man more aquatic in the past?', *New Scientist*, **7,** 642–5.

35 HARLOW, H. F. and R. R. ZIMMERMANN (1959). 'Affectional responses in the infant monkey', *Science, N.Y.*, **130,** 431.

36 HINDE, J. S. and R. A. HINDE (eds.) (1973). *Constraints on Learning; Limitations and Predispositions*. London and New York, Academic Press.

37 HINDE, R. A. (1970). *Animal Behaviour; a Synthesis of Ethology and Comparative Psychology*, 2nd edn. New York, McGraw-Hill.

38 HOLT, J. (1970). *How Children Fail*. Harmondsworth, Penguin Books.

39 HOLT, J. (1971). *How Children Learn*. Harmondsworth, Penguin Books.

40 HUTT, C. and S. J. HUTT (1970). *Direct Observation and Measurement of Behavior*. Springfield, Illinois, Charles C. Thomas.

41 HUXLEY, J. (1948). *Man in the Modern World*. New York, Mentor Books.

42 KANNER, L. (1943). 'Autistic disturbances of affective contact', *Nerv. Child*, **2**, 217–50.

43 KETTLEWELL, H. B. D. (1955). 'Selection experiments on industrial melanism in the Lepidoptera', *Heredity*, **9**, 323–42.

44 KETTLEWELL, H. B. D. (1956). 'Further selection experiments on industrial melanism in the Lepidoptera', *Heredity*, **10**, 287–301.

45 KONISHI, M. and F. NOTTEBOHM (1969). 'Experimental studies in the ontogeny of avian vocalizations', in *Bird Vocalizations in Relation to Current Problems in Biology and Psychology*, ed. R. A. Hinde. London, Cambridge University Press.

46 KREBS, J. R., M. H. MACROBERTS and J. M. CULLEN (1972). 'Flocking and feeding in the Great Tit; an experimental study', *Ibis*, **114**, 507–31.

47 KRUUK, H. (1964). 'Predators and anti-predator behaviour of the Black-headed Gull (*Larus ridibundus*)', *Behaviour Suppl.*, **11**.

48 KRUUK, H. (1972). *The Spotted Hyena*. Chicago, University Press.

49 KUMMER, H. (1968). *The Social Organisation of Hamadryas Baboons*. Basel, S. Karger.

50 LACK, D. (1968). *Ecological Adaptations for Breeding in Birds*. London, Methuen.

51 LEE, R. B. and I. DE VORE (eds.) (1968). *Man the Hunter*. Chicago, Aldine Publishing Company.

52 LEHRMAN, D. S. (1970). 'Semantic and conceptual issues in the nature-nurture problem', in *Development and Evolution of Behavior*, ed. L. R. Aronson, E. Tobach, D. S. Lehrman and J. S. Rosenblatt. San Francisco, Freeman.

53 LORENZ, K. (1965). *Evolution and Modification of Behavior*. Chicago, University Press.

54 MCGREW, W. C. (1972). *An Ethological Study of Children's Behavior*. London and New York, Academic Press.

55 MACLANNAHAN, H. (1970). 'Studies of behaviour ontogeny in gulls'. D.Phil. thesis Oxford. (*Behaviour* in the Press.)

56 MANNING, A. (1956). 'Some aspects of the foraging behaviour of bumblebees', *Behaviour*, **9**, 164–203.

57 MARLER, P. (1967). 'Comparative study of song development in sparrows', *Proc. Int. Orn. Congr.*, **14**, 231–44.

58 MAYR, E. (1963). *Animal Species and Evolution*. Cambridge, Mass., Harvard University Press.

59 MECH, L. D. (1966). 'The wolves of Isle Royale', *U.S. Nat. Park Serv. Fauna Ser.*, **7**, 1–210.

60 MECH, L. D. (1970). *The Wolf*. London, Constable.

61 MEDAWAR, P. B. (1959). *The Future of Man*. New York, Mentor Books.

62 MORRIS, D. (1967). *The Naked Ape*. London, Jonathan Cape.

63 NELSON, J. B. (1965). 'The behaviour of the gannet', *Br. Birds*, **58**, 233–88, 313–36.

64 NORTON-GRIFFITHS, M. (1967). 'Some ecological aspects of the feeding behaviour of the Oystercatcher on the Edible Mussel', *Ibid*, **109**, 412–24.

65 NORTON-GRIFFITHS, M. (1968). 'The feeding behaviour of the Oystercatcher (*Haematopus ostralegus*)', D.Phil. thesis, Oxford, and in preparation.

66 O'GORMAN, G. (1970). *The Nature of Childhood Autism*. London, Butterworth.

67 PATTERSON, I. J. (1965). 'Timing and spacing of broods in the Black-headed Gull *Larus ridibundus* L.', *Ibis*, **107**, 433–60.

68 RIBBANDS, C. R. (1953). *The Behaviour and Social Life of Honeybees*. London, Bee Research Association.

69 ROEDER, K. D., A. E. TREAT and J. S. VANDEBERG (1968). 'Auditory sense in certain sphingid moths', *Science, N.Y.*, **159**, 331–3.

70 SCHALLER, G. B. (1972). 'Predators of the Serengeti', *Nat. Hist., N.Y.*, **81**, 38–49, 60–69.

71 SCHALLER, G. B. (1972). *The Serengeti Lion*. Chicago, University Press.

72 SCHALLER, G. B. and G. R. LOWTHER (1969). 'The relevance of carnivore behaviour to the study of early hominids', *SWest. J. Anthrop.*, **25**, 307–41.

73 SELIGMAN, M. E. P. (1970). 'On the generality of the laws of learning', *Psychol. Rev.*, **77**, 406–18.

74 SIMMONS, K. E. L. (1966). 'Anting and the problem of self-stimulation', *J. Zool., Lond.*, **149**, 145–62.

75 SMITH, J. N. M. (1971). 'Studies of the searching behaviour and prey recognition of certain vertebrate predators'. D.Phil. thesis, Oxford, and in preparation.

76 SMITH, N. G. (1966). 'Adaptations to cliff-nesting in some arctic gulls (*Larus*)', *Ibis*, **108**, 68–83.

77 SPENCER-BOOTH, Y. and R. A. HINDE (1971). 'Effects of brief separations from mothers during infancy on behaviour of Rhesus monkeys 6–24 months later', *J. Child Psychol. Psychiat.*, **12**, 157–72.

78 THORPE, W. H. (1961). *Bird Song*. London, Cambridge University Press.

79 TIGER, L. (1969). *Men in Groups*. New York, Random House.

80 TINBERGEN, E. A. and N. TINBERGEN (1972). 'Early childhood autism—an ethological approach', *Z. Tierpsychol. Suppl.*, **10**, 1–53. Berlin: Parey.

81 TINBERGEN, L. (1960). 'The natural control of insects in pine woods', *Arch. néerl. Zool.*, **13**, 265–379.

82 TINBERGEN, N. (1958). *Curious Naturalists*. London, Country Life.

83 TINBERGEN, N. (1964). 'On adaptive radiation in gulls (tribe Larini)'. *Zool. Med.*, **39**, 209–23.

84 TINBERGEN, N. (1965). 'Von den Vorratskammern des Rotfuchses (*Vulpes vulpes* L.)'. *Z. Tierpsychol.*, **22**, 119–49. (Translated in Vol. 1 of *The Animal in its World*, p. 315.)

85 TINBERGEN, N. (1967). 'Adaptive features of the Black-headed Gull.' *Proc. Int. Orn. Congr.*, **14**, 43–59.

86 TINBERGEN, N. (1973) *The Animal in its World*. Vol. 1. *Field Studies*. London, Allen and Unwin.

87 TINBERGEN, N., G. J. BROEKHUYSEN, F. FEEKES, J. C. W. HOUGHTON, H. KRUUK AND E. SZULC (1962). 'Egg shell removal by the Black-headed Gull, *Larus ridibundus* L.; a behaviour component of camouflage', *Behaviour*, **19**, 74–117. (Reprinted in Vol. 1 of *The Animal in its World*, p. 250.)

88 TINBERGEN, N., M. IMPEKOVEN and D. FRANCK (1967). 'An experiment on spacing-out as a defence against predation', *Behaviour*, **28**, 307–21. (Reprinted in Vol. 1 of *The Animal in its World*, p. 329.)

89 TINBERGEN, N., H. KRUUK, M. PAILLETTE and R. STAMM (1962). 'How do Black-headed Gulls distinguish between eggs and egg shells?', *Br. Birds*, **55,** 120–9. (Reprinted in Vol. 1 of *The Animal in its World*, p. 304.)

90 TSCHANZ, B. (1968). 'Trottellummen', *Z. Tierpsychol. Suppl.*, **4,** 1–103.

91 UEXKÜLL, J. VON (1921). *Umwelt und Innenwelt der Tiere*. Berlin, Springer.

92 WARD, B and R. DUBOS (1972). *Only one Earth*. London, Deutsch.

93 WASHBURN, S. L. and C. S. LANCASTER (1968). 'The evolution of hunting', in *Man the Hunter*, ed. R. B. Lee and I. DeVore, pp. 293–304. Chicago, Aldine Publishing Company.

94 WHITE, S. J. (1971). 'Selective responsiveness by the Gannet (*Sula bassana*) to played-back calls', *Anim. Behav.*, **19,** 125–31.

95 WICKLER, W. (1967). 'Socio-sexual signals and their intra-specific imitation among Primates', in *Primate Ethology*, ed. D. Morris, pp. 69–148. London, Weidenfeld and Nicolson.

96 WYNNE-EDWARDS, V. C. (1962). *Animal Dispersion in Relation to Social Behaviour*. Edinburgh, Oliver & Boyd.